SUSTAINABILITY, CITIZEN PARTICIPATION, AND CITY GOVERNANCE

Multidisciplinary Perspectives

Edited by Hoi L. Kong and Tanya Monforte

The inaction of nation states and international bodies has posed significant risks to the environment. By contrast, cities are sites of action and innovation. In *Sustainability, Citizen Participation, and City Governance*, contributors researching in the areas of law, urban planning, geography, and philosophy identify approaches for tackling many of the most challenging environmental problems facing cities today.

Sustainability, Citizen Participation, and City Governance facilitates two strands of dialogue about climate change. First, it integrates legal perspectives into policy debates about urban sustainability and governance, from which law has typically stood apart. Second, it brings case studies from Quebec into a rare conversation with examples drawn from elsewhere in Canada.

The collection proposes humane and inclusive processes for arriving at effective policy outcomes. Some chapters examine governance mechanisms that reconcile clashes of incommensurable values and resolve conflicts about collective interests. Other chapters provide platforms for social movements that have faced obstacles to communicating to a broad public. The collection's proposals respond to drastic changes in urban environments. Some changes are imminent. Others are upon us already. All threaten the present and future well-being of urban communities.

HOI L. KONG holds the Rt. Hon. Beverley McLachlin, P.C., UBC Professorship in Constitutional Law at the Peter A. Allard School of Law at the University of British Columbia.

TANYA MONFORTE is an assistant professor of political science at Concordia University.

Sustainability, Citizen Participation, and City Governance

Multidisciplinary Perspectives

EDITED BY HOI L. KONG AND TANYA MONFORTE

UNIVERSITY OF TORONTO PRESS
Toronto Buffalo London

© University of Toronto Press 2022
Toronto Buffalo London
utorontopress.com

ISBN 978-1-4875-4297-9 (cloth)   ISBN 978-1-4875-4299-3 (EPUB)
ISBN 978-1-4875-4298-6 (paper)   ISBN 978-1-4875-4300-6 (PDF)

Library and Archives Canada Cataloguing in Publication

Title: Sustainability, citizen participation, and city governance : multidisciplinary perspectives / edited by Hoi L. Kong and Tanya Monforte.
Names: Kong, Hoi L., editor. | Monforte, Tanya, editor.
Description: Includes bibliographical references and index.
Identifiers: Canadiana (print) 20220243778 | Canadiana (ebook) 20220243875 | ISBN 9781487542986 (paper) | ISBN 9781487542979 (cloth) | ISBN 9781487542993 (EPUB) | ISBN 9781487543006 (PDF)
Subjects: LCSH: Sustainable urban development – Canada – Citizen participation – Case studies. | LCSH: Urbanization – Environmental aspects – Canada – Case studies. | LCSH: City planning – Canada – Case studies. | LCGFT: Case studies.
Classification: LCC HT243.C3 S87 2022 | DDC 307.760971 – dc23

We wish to acknowledge the land on which the University of Toronto Press operates. This land is the traditional territory of the Wendat, the Anishnaabeg, the Haudenosaunee, the Métis, and the Mississaugas of the Credit First Nation.

The publication of this book was made possible by support from the Centre for Interdisciplinary Research on Montreal, the Peter Wall Institute for Advanced Studies at UBC, the Social Sciences and Humanities Research Council of Canada, and the McConnell Foundation.

University of Toronto Press acknowledges the financial support of the Government of Canada, the Canada Council for the Arts, and the Ontario Arts Council, an agency of the Government of Ontario, for its publishing activities.

 Canada Council for the Arts   Conseil des Arts du Canada

 ONTARIO ARTS COUNCIL
CONSEIL DES ARTS DE L'ONTARIO
an Ontario government agency
un organisme du gouvernement de l'Ontario

Funded by the Government of Canada   Financé par le gouvernement du Canada

Contents

Introduction 3
HOI L. KONG AND TANYA MONFORTE

Part One: Social Movements and Innovation

1 Beyond Smart/Sustainable Cities: Towards a Citizen-Centric Rebel Cities Transition 13
JONATHAN DURAND FOLCO

2 Mobilisons-Nous: "Violent Infrastructure" and Pedestrian Space in Montreal 44
KEVIN MANAUGH AND NATALYA BEREZINA DRESZER

3 Boroughs, Small Municipalities, and Sustainability: What Is Municipal Innovation and Can It Make a Difference? 65
RICHARD SHEARMUR

Part Two: The Role of Law and Overcoming Collective Action Problems

4 Sustainable Development and Property Rights: Citizen Participation in Dismantling Urban Environmental Regulation in British Columbia 95
DEBORAH CURRAN

5 Sustainable Urban Design: The Case of Montreal 118
HOI L. KONG

6 The Implications of Stakeholder Group Involvement in Urban Sustainable Development 137
ALEXANDRA FLYNN

7 Complementing Citizen Engagement with Innovative Forms of Professional Co-production: A Renewed Case for Transdisciplinary Charrettes   163
NIK LUKA, BRIA AIRD, AND NINA-MARIE LISTER

Afterword: Thinking Through Transdisciplinarity in Urban Sustainability   194
TANYA MONFORTE

Contributors   217

Index   221

SUSTAINABILITY, CITIZEN PARTICIPATION, AND CITY GOVERNANCE

Multidisciplinary Perspectives

Introduction

HOI L. KONG AND TANYA MONFORTE

More than nation-states, cities will be forced into the frontlines by global warming, energy and water insecurity, and other environmental challenges.

Saskia Sassen (2009)

There are significant risks to the environment posed by the inaction of nation states and international bodies. But citizens seeking policy change may view national and international organizations as too remote to engage effectively and their legal and policy frameworks too complex to navigate. By contrast, city governments are close to the lived experience of all who reside within urban boundaries. And so, because their legal and policy processes are relatively accessible to city dwellers, cities hold out the promise of effective responses to the environmental challenges that Sassen identifies in the epigraph above. As Yishai Blank has noted, the inherent complexity of regulating environmental issues is made more manageable when viewed through the lens of urban law and policy (Blank, 2006). Moreover, although cities are relatively small political units, when the aggregated environmental impact of city dwellers' activities across the planet are considered, the global significance of sustainable urban governance becomes evident. There is therefore a growing consensus that "where states fail, local governments might succeed" (Blank, 2006, p. 918), and a growing body of literature on urban sustainable development.

Before we identify the contributions of this collection to the literature, we first define what we mean by sustainable development. Hoi Kong and Alexandra Flynn draw out different elements of this concept's definition, as it is set out by the World Commission on Environment and Development (1987). In Chapter 5, Kong writes that sustainable development that "aims to meet present needs should not compromise the environmental conditions necessary for future generations to meet their needs." In Chapter 6, Flynn fleshes out this definition, with the condition, standard in the literature, that "sustainability must be based on a balance between social equity, environmental protection, and economic growth." Each chapter of this collection highlights the intergenerational aspect of sustainable development and/or the policy aspiration of balancing social, environmental, and economic goals.

We turn now to identify three contributions of this collection to the literature on urban sustainable development.

The first is its distinctive approach to urban sustainable development, drawn from case studies in Quebec and the rest of Canada. It recognizes the need for policy that is inclusive and reflective, in the area of urban sustainable development. The collection's approach to inclusivity is inspired by the experience and expertise of the diverse, multicultural urban populations in Quebec and across Canada. And in recognition of the complexity and mutability of the environmental challenges these populations face, the approach is reflexive: it prescribes institutions and policy instruments that can adapt to multifaceted and unforeseen conditions.

Second, this collection identifies and fills a gap in English-language literature on urban sustainable development. Case studies and examples from Quebec are under-represented in this body of writing, and in response to this absence, this collection brings the Quebec (and especially Montreal), experience into conversation with select urban sustainable development initiatives in other parts of Canada.[1] The particular "history of social movements and radical urban activism in Montreal" (McMahon & Oddie, 2007) distinguishes it from the rest of North America. This distinctive tradition of activism drives social change in Quebec's cities, including in the area of sustainable development. Several texts in this collection make the Quebec experience available to an anglophone readership.

Third, this collection brings together legal scholars and academics from geography, architecture, and environmental studies. Although urban studies is a deeply interdisciplinary field, legal scholars and legal practitioners have been absent from the field in Canada, with a few exceptions. This collection seeks to integrate legal perspectives more fully into one subset of the field: Canadian debates about urban sustainability and governance.

Robinson and Dale's key insight (2012) – that "government gridlock ... is a major barrier" to realizing urban sustainable development goals provides a point of entry for the dialogue between law and other disciplines that make up the field of urban studies in Canada. Several authors in

---

1 It is difficult to demonstrate an absence in the literature definitively, but we take as examples edited collections in urban sustainability such as Dale et al. (2012), Toner and Meadowcroft (2009), and Robinson et al. (2014).

this collection highlight how clashing interests within cities can frustrate attempts to shift law and policy in this area. Other authors assess the role that social movements can play in overcoming gridlock and achieving sustainable development objectives.

These areas of focus bring to light the significant, though understudied, functions that law can play as cities pursue sustainable development objectives. Several authors in this collection examine how law, made and applied in cities, can facilitate a green and inclusive turn in policy and governance. In general, the authors adopt a pragmatic stance. They recognize that urban law and policy can mediate clashes between competing interests but can also contribute to political gridlock. Several chapters evince this pragmatism, as they address the limited role that law and policy processes can play in resolving gridlock within select cities in Quebec and the rest of Canada.

In addition to being pragmatic, the approach to urban sustainable development that we draw from the case studies is reasoned and innovative. The policy proposals that the collection's authors identify are reasoned, in that they are supported by experience and evidence, and are often justified in those terms by governments and popular movements alike. The approach is innovative, in the colloquial sense that it pursues new ideas or methods (we shall see below that "innovation" can also be defined in more academic terms). In particular, the approach embraces insights and initiatives emerging from beyond the state and its standard interlocutors. And when proponents of the approach do look within the agencies of the state, they identify measures that destabilize the status quo.

This pragmatic, reasoned, and innovative approach to policy innovation is evident in how the collection's authors tackle difficult collective action problems. Although several important texts have made the normative case for sustainability and set out plans to realize sustainable development goals in the urban context (Robinson et al., 2014), authors do not always directly address the challenges posed by profound disagreement and conflicting interests. Several contributors to this collection examine decision-making processes that respond to coordination and assurance problems, which make cooperation in the public interest difficult.[2] Their chapters examine how – by facilitating inclusive and grounded discussion about policy responses to pressing environmental concerns – governments address problems such as NIMBYism. Other authors examine how, in the context of sustainable development initiatives, divergent

---

2 For an analysis of this coordination problem, see Mayer (2014).

interests in urban polities can directly clash. Among the conflicts considered are those between motorists on the one hand, and cyclists and pedestrians on the other; between individual businesses and communities; and between private property rights and collective interests.

Perhaps reflecting the deepest aspirations of the multicultural communities from which their case studies are drawn, the collection's authors propose humane processes. Ideally, such processes enable urban polities to arrive at effective policy outcomes, in the face of deep disagreement about matters of value and about the very nature and significance of collective interests. Yet although the collection focuses on processes and institutions, its ambitions are not simply technical or technocratic. In urban sustainable development, the moral claims of future generations are compelling and constant. So too are the need to reconcile diverse interests in ways that are reasonable and reasoned, and the imperative to give voice to populations who have been marginalized for too long. This collection's aims are therefore normative. Its proposals respond to drastic changes in urban environments that are imminent or upon us already and that threaten the present and future well-being of our communities. The collection examines governance mechanisms that seek to reconcile sometimes incommensurable values, and gives voice to some (but of course not all) social movements that have faced obstacles to making their insights available to a broad public.[3]

With these normative stakes clearly in view, we now provide an overview of the chapters.

Part I: Social Movements and Innovation

In Part I, the authors address social movements and urban innovation.

The first set of chapters foreground social movements, provide perspectives on the meaning of empowerment, and offer strategies to achieve it in urban sustainability initiatives. Jonathan Durand Folco's "Beyond Smart/Sustainable Cities: Towards a Citizen-Centric Rebel Cities Transition" explores the forms smart cities can take. The concept of the smart city is presented as a way to reconcile the requirements of sustainability initiatives, strategies, and governance, on the one hand, and

---

3 The absence of Indigenous perspectives on environmental sustainability is a shortcoming of the collection. The original meeting from which this collection sprang did not have a territorial tie to North America. The question arose about how to best incorporate Indigenous perspectives after the collection veered in the editing stage towards a located specialization. Flagging this shortcoming, we hope, will spur the publishers to seek out and publish further in this important area.

demands for citizen participation, on the other. The chapter sketches out three different models of smart cities and explains how each has implications for sustainability and citizen participation. Durand Folco shows how each of the techno-centric neoliberal, collaborative, and citizen-centric smart cities provides a context that gives the concept of "empowerment" meaning. He argues that depending on the model of the smart city, empowerment can have a collective meaning, an individual sense, or disempowering effects. He argues against the tendency, prevalent in the literature, of collapsing the concepts of sustainability and the smart city, and he ultimately champions the "rebel city" as the model that best meets sustainability objectives.

In "Mobilisons-Nous: 'Violent Infrastructure' and Pedestrian Space in Montreal," Kevin Manaugh and Natalya Berezina Dreszer provide case studies in which residents struggle to make Montreal's streets more accessible and inclusive. According to the authors, achieving environmental and social sustainability in transportation policy requires that streets be organized differently, as public spaces. Manaugh and Berezina Dreszer examine political struggles in Montreal in which participants have aimed to make pedestrians, rather than motorists, the primary social units of a green city. In so doing, they offer lessons in how to empower social movements (in Montreal and elsewhere) to make their cities more pedestrian-oriented and therefore more sustainable.

In the second section of Part I, Richard Shearmur presents a standard but contested view of innovation. In the face of impending and lasting environmental harms and sometimes sclerotic institutions, we may naturally seek out policy innovation as an end in itself, without being cognizant of its potential costs or examining whether specific innovations are effective in achieving policy goals. Shearmur's "Boroughs, Small Municipalities, and Sustainability: What Is Municipal Innovation and Can It Make a Difference?" examines potential costs of urban policy innovation, but argues that they may be minimized in smaller municipalities. Shearmur examines policy innovations in sustainable development in several Quebec municipalities and notes that policy innovation is always associated with potential costs. However, because the municipalities he examines are small, relative to major metropolitan centres, provinces, and nation states, they can experiment without large-scale harm. Therefore the potential costs associated with such innovation are relatively low. Shearmur further argues that we can learn which urban innovations work in this domain by analysing municipal networks. Finally, he notes that because of the sheer number of municipal governments, successful urban initiatives can, when taken in the aggregate, contribute to progress towards a state's goals in sustainable development.

Part II: The Role of Law and Overcoming Collective Action Problems

The group of chapters in this part tackle collective action problems that arise when cities in nations with advanced consumption economies transition to sustainable policy agendas (Lafferty & Meadowcroft, 2000; Toner & Meadowcroft, 2009). In particular, these chapters address conflicts between individual interests and collective goods. Such collective action problems can arise when the transition from a car-centred to a pedestrian-centred transportation system involves removing parking spaces from the streets. In this example, the collective goods associated with such a transition, including reductions in greenhouse gas emissions, confront the interests that some individuals have in convenient parking. The authors in this part examine the role that law can play in addressing collective action problems associated with urban sustainable development, and they assess the extent to which planning can be made inclusive and effective.

Deborah Curran's "Sustainable Development and Property Rights: Citizen Participation in Dismantling Urban Environmental Regulation in British Columbia" provides a case study in what happens when sustainable land use regulation and private property rights are presented as being in conflict. Curran explains that in Canada there is no constitutional or legal requirement to compensate owners for changes in property values due to environmental regulation. Nonetheless, in the course of participatory processes in British Columbia, residents and property owners advanced property rights and arguments that sought the repeal of environmental land use protections. Curran's chapter is a cautionary tale for advocates who may be overly optimistic about the potential for citizen participation to yield sustainable urban development outcomes.

In "Sustainable Urban Design: The Case of Montreal," Hoi Kong analyses situations in which the city functions as a commons whose resources are at risk of being depleted. He argues that deliberative democratic theory can inform how we understand the concepts of urban sustainability and the city as a commons, and he claims that the theory can guide legal reforms. His case study focuses on Montreal's recent consultation on reducing fossil fuel dependence, and he examines how legal frameworks can mitigate problems of democratic local governance. His deliberative democratic analysis reveals how well-structured consultations can overcome the dominance of special interest groups and address the exclusion of future generations from present democratic processes.

In "The Implications of Stakeholder Group Involvement in Urban Sustainable Development," Alexandra Flynn delves into the conflicting agendas of interest groups, as she examines how local businesses, neighbourhood associations, developers, and residents interact in Toronto's urban governance processes. The chapter explores how social, economic, and environmental interests can conflict when sustainable development is put into practice. Finally, Flynn explores how interest groups (especially neighbourhood associations) can influence governance, and she assesses how ad hoc decision-making responds to policy challenges.

The collection concludes on a methodological note about the importance of transdisciplinary collaboration in sustainable urban development. In "Complementing Citizen Engagement with Innovative Forms of Professional Co-production: A Renewed Case for Transdisciplinary Charrettes," Nik Luka, Bria Aird, and Nina-Marie Lister argue that as urban planners help to develop more sustainable modes of urban design, transdisciplinary collaboration is becoming an important aspect of planning practice. The authors use the charrette as a model of productive transdisciplinary coproduction. A charrette is an intense workshop, often used by architects, urban planners, and designers, that allows projects to be produced and delivered quickly. This "compression of space and time" creates a sense of urgency that focuses participants' attention on the problem at hand and enables them to arrive at concrete solutions, despite disciplinary differences among participating specialists. The authors offer pragmatic suggestions for ways to make transdisciplinary charrettes work, in the context of sustainable urban design. Perhaps most provocatively, they argue that there are moments in planning processes when professional perspectives should be given prominence and perhaps priority over the insights of the general public.

Finally, in the afterword, Tanya Monforte draws together the themes of the collection and suggests how bridging differences between disciplines can yield insights and generate new policy possibilities. With an eye to where the urban sustainable development conversation should go next, she identifies models of knowledge production that can help scholars and practitioners grapple with problems in this context. She further argues that as transdisciplinary methods become more entrenched in sustainability studies, we should ensure that research methods remain flexible, reflexive, and inclusive. She argues that transdisciplinary methods ultimately provide us with opportunities to imagine different ways to collaborate and think through complex, urgent, and potentially existential social problems.

REFERENCES

Blank, Y. (2006). The city and the world. *Columbia Journal of Transnational Law*, *44*(3), 875–939. 10.4324/9781315092485-7

Buxton, G. V. (1992). Sustainable development and the summit: A Canadian perspective on progress. *International Journal*, *47*(4), 776–795. https://doi.org/10.2307/40202803

Dale, A., Dushenko, W. T., & Robinson, P. (Eds.). (2012). *Urban sustainability: Reconnecting space and place.* University of Toronto Press.

Lafferty, W. M., & Meadowcroft, J. (2000). Introduction. In W. M. Lafferty & J. Meadowcroft (Eds.), *Implementing sustainable development: Strategies and initiatives in high consumption societies* (pp. 1–22). Oxford University Press. 10.1093/0199242011.003.0001

Mayer, F. W. (2014). *Narrative politics: Stories and collective action.* Oxford University Press.

McMahon, M., & Oddie, R. (2007). Urban sustainability and environmental research in Canada: Prospects for overcoming disciplinary divides. https://www.yorku.ca/cityinstitute/projects/projects/urban-sustainability-and-environmental-research-in-canada-prospects-for-overcoming-disciplinary-divides/

Robinson, J. B., Francis, G., Legge, R., Lerner, S., & Slocomve, D. S. (2014). *Life in 2030: Exploring a sustainable future for Canada: Sustainability and the environment.* University of British Columbia Press.

Sassen, S. (2009). Cities are at the center of our environmental future. *Sapiens,* 2(3), 1–8. http://sapiens.revues.org/948

Toner, G., & Meadowcroft, J. (2009). *Innovation, science, environment 1987–2007.* McGill-Queen's University Press.

World Commission on Environment and Development. (1987). *Report of the World Commission on Environment and Development: Our common future.* World Commission on Environment and Development.

PART ONE

Social Movements and Innovation

1 Beyond Smart/Sustainable Cities: Towards a Citizen-Centric Rebel Cities Transition

JONATHAN DURAND FOLCO

Introduction

In order to analyse how citizens and other stakeholders formulate sustainable urban development, it is essential to assess the notion of the "smart city" and its meanings in public debate. The smart city quickly became a buzzword used by a wide range of actors (businesses, consulting firms, architects, engineers, scholars, community organizations, local politicians, and real-estate developers) to promote digital technologies to improve resource management, innovation, governance, mobility, and quality of life. But the "smart" craze is more than a temporary fashion; it is framing economic development strategies, sustainability policies, and the way people understand their relationship to the city. If digital technologies are becoming an increasingly important mediation between sustainability initiatives and citizen participation in the city level, likely the "smartness" discourse will have a strong influence on local development, social interactions, and the urban environment as a whole. Hence, "code is mediating the various parts of city life" (Amin & Thrift, 2002, p. 125).

Nonetheless, there is still no consensus on the exact meaning of smart cities, as authors identified 23 definitions in recent literature (Albino et al., 2015). Furthermore, we see a rapid proliferation of terms and models aiming to redefine urban development across several dimensions, thus creating blurry frontiers between sustainable–smart–resilient–low carbon–eco–knowledge cities (de Jong et al., 2015). More recently, the overlapping research on sustainable cities and smart cities converged in the idea of "smart sustainable cities" (Freeman, 2017; Ibrahim et al., 2018), which provides an aggregated concept and integrated urban vision (Höjer & Wangel, 2015). Some scholars emphasize the role of big data and digital technologies for advancing sustainability in urban contexts (Bibri, 2018), while other interpretations insist on new opportunities for multi-stakeholder participation and collaborative governance (Pereira et al., 2017).

The attempt to integrate new technologies, sustainability, and participation in a comprehensive framework is not free of tensions, conflicts, or even contradictions. The objective of this chapter is to show that there is not one unified theory, but three broad conceptions of smart cities, and that each model implies a different perspective on sustainability and citizen participation. Using discourse analysis, theories of public participation, and recent research on smart citizenship, we distinguish three different and perhaps conflicting models of the city: (1) the techno-centric neoliberal smart city; (2) the collaborative sustainable city; and (3) the citizen-centric rebel city.

This threefold typology is inspired by several authors who analysed the different meanings of empowerment (Bacqué & Biewener, 2013), environmental discourses (Dryzek & Schlosberg, 2005), and smart cities (Kitchin, 2016; Cliche, 2017). The ideal-type approach used to build these models rests on three presuppositions. First, discourses are embedded in political, economic, and historical contexts. Second, discourses are social constructions enmeshed in dynamic power relationships among actors, institutions, and structures of domination. Third, one concept can be given different content through "chains of equivalences," the signification of a notion being constituted by the images, symbols, and propositions with which it is related (Laclau & Mouffe, 2001). In other words, "smart city" will not have the same meaning if it is associated primarily with economic growth or sustainable development, technological innovation or grassroots innovations, individual preferences or collective needs, entrepreneurship or social emancipation.

Before analysing each model of the city separately, it is necessary to have a closer look to several smart city definitions to understand their key dimensions and see how they are related to sustainability and participation. This first step is important, because the smart city hype could easily undermine normatively compelling understanding of sustainability by exacerbating social and economic inequalities, and diverting investment from public infrastructures towards privatized modes of service, such as Uber, that favour corporations at the expense of the public interest. Furthermore, a too narrow understanding of "participation" could dismiss more transformative modes of citizen engagement.

After sketching the first two models from a descriptive and critical point of view, we will formulate a normative argument to emphasize the need to promote the citizen-centric rebel cities approach as an alternative to the "smart" and "sustainable" cities discourse that currently dominates the public urban development debate. While the smart neoliberal city ignores social justice issues and defends a "technological solutionism" approach to solve environmental problems (Morozov,

2013), the sustainable cities discourse does not go far enough to grapple with social, ecological, and democratic challenges in a period of *global crisis* (Fraser, 2017). Hence, we believe that the smart and sustainable cities frameworks propose a weak form of citizen participation, are insufficiently critical of power relations between unequal actors in complex governance, and are incapable of proposing a convincing political project to answer the social, political, and ecological challenges of the twenty-first century.

The Multiple Dimensions of Smart Cities

If the term "smart" has been used since the 1990s, alongside the Smart Growth and New Urbanism movements, which seek to promote urban densification with transit-oriented and mixed-use development (Ouellet, 2006), it is now associated mainly with the integration of information and communications technologies (ICT) into urban infrastructures: sensors, screens, big data, mobile broadband, algorithms, cloud computing, cameras, mobile phones, Internet of Things, etc. The word "smart" was used mainly at the beginning of the twenty-first century "as an 'urban labelling' phenomenon" (Freeman, 2017, p. 27), and it is now the most popular city label around the world (de Jong et al., 2015). The smart city is supported by organizations such as think tanks, research institutes, and government initiatives, and some countries adopted this perspective on a massive scale, as India announced its plan to create more than 100 smart cities to respond to increasing urban challenges (Smart Cities India, 2015).

Far from being reduced to digital technologies in urban infrastructures, the smart city has a wide range of applications in domains like government, security and emergency services, transportation, energy, waste, environment, and buildings and homes (Kitchin, 2016). The "smart city wheel" in Figure 1.1 shows several dimensions connected to this notion. To take another example, the European Smart Cities initiative identified six key dimensions: smart governance, smart people, smart mobility, smart economy, smart environment, and smart living (Giffinger et al., 2007). Each dimension can be divided into different indicators, ranking systems, and other measures to assess smart cities performance.

However, most definitions of smart cities put unequal emphasis on those several dimensions. Digital technologies always play a central role, and even if they are supposed to solve numerous problems in various spheres, sustainability and citizen participation are often considered as secondary issues. As Haarstad has shown with extensive empirical data, "the smartness approach is strongly tied to innovation, technology, and economic entrepreneurialism, and sustainability does not appear to be a very important

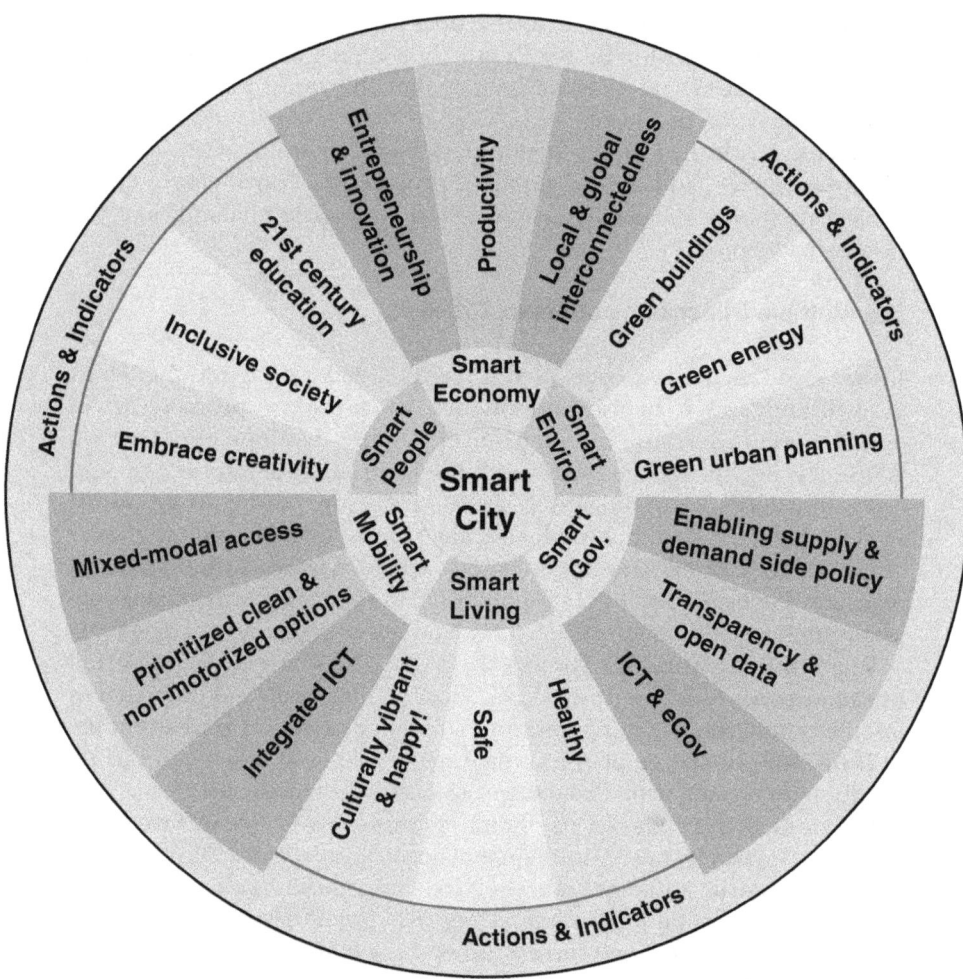

Figure 1.1 Smart city wheel. Image courtesy of Boyd Cohen.

motivating driver" (Haarstad, 2016, p. 423). We can then affirm that smart cities have no internal or conceptual relation to sustainability and participation, although those dimensions can be mediated by technology. Overall, smart refers "to any advanced technology deployed in cities with the intent of optimizing the use of resources, producing new resources, changing user behaviour, or promising other kinds of gains in terms of, for example, flexibility, security, and sustainability" (Morozov & Bria, 2018, p. 4).

The Emergence of the Techno-Centric Neoliberal Smart City

The first step to understand the prevailing logic of the smart city is to explore the connection between techno-centric approaches and the influence of neoliberalism on urban development. Our argument is not that the smart city is necessarily neoliberal, or that modern forms of capitalist urban development always take the form of techno-centrism. In other words, the assumptions that smart cities are systematically neoliberal and techno-centric is presumptive and should be challenged. For example, we can observe many techno-centric socialist cities in China today. But it is still useful to analyse the meeting point between techno-centric and neoliberal discourses in smart cities, because these two logics usually go hand in hand.

The first and most widespread discourse on smart cities is techno-centric, meaning that it relies mainly on the transformative power of digital technologies, computing, and big data.

> These data-driven, networked technologies work to make cities knowable and controllable in new, dynamic, reactive ways through the use of vast quantities of real-time data and interactive, programmable systems. Moreover, the data generated can be used to create and improve models and simulations to guide future urban development. (Kitchin, 2016, p. 11)

In this view, environmental issues are treated as problems that could be fixed by technological innovations. For example, "strategic smart sustainable urban planning" is based on "big data analytics and context-aware computing" (Bibri, 2018, p. vii). This smart city model is closely related to other discourses, like the techno-centric conception of ecological transition (Audet, 2016), solutionism (Morozov, 2013), data-driven adaptive management (Akhtar et al., 2016), and ecological modernization (Hajer, 1995). These discourses can be related to the "Promethean response" to the ecological crisis, referring to the Greek god Prometheus who stole the fire from Zeus and had unlimited confidence in the human capacity to manipulate the world and overcome any problem through the use of technology (Dryzek & Schlosberg, 2005).

The two best examples of techno-centric smart cities are Masdar, in United Arab Emirates, and the Songdo International Business District, in South Korea. These two massive urban projects built from scratch incorporate the most advanced technologies in terms of renewable energies, LEED-certified buildings, personal rapid transit, pneumatic waste disposal system, etc. Songdo is conceived as a "ubiquitous city," that is an urban system where "every device, component, and service within the city is linked to an information network, largely through wireless networking channels" (Cohen, 2014). This "real-time city" and smart urbanism raise ethical issues, like "the politics of big urban data, technocratic governance and city development, corporatisation of city governance and technological lock-ins, buggy, brittle and hackable cities, and the panoptic city" (Kitchin, 2013, p. 1).

Techno-centric smart cities favour data-driven management, public-private partnerships, and big corporations running digital urban infrastructures, like IBM, Cisco, or Alphabet. For example, Toronto was recently "Google's guinea-pig city" (Sauter, 2018) with the Sidewalk Toronto project. This project, which started in 2017 and was abandoned in 2020, was a partnership between Waterfront Toronto and Alphabet's Sidewalk Labs aiming to build "a new kind of mixed-use, complete community [combining] urban design and new digital technology to create people-centred neighbourhoods that achieve precedent-setting levels of sustainability, affordability, mobility, and economic opportunity" (Sidewalk Labs, 2018).[1] If this discourse looks harmless on the surface, this urban development model based on technological solutionism is strongly put forward by market actors, big corporations like Cisco, IBM, Google, Microsoft, and "sharing economy" private companies like Uber and Airbnb, that promote neoliberalism (Grossi & Pianezzi, 2017).

Neoliberalism is often conceived as an ideology or economic policy aiming to dismantle social provision, state intervention, and wealth redistribution. Neoliberalism also promotes private investments as the main driver of economic growth, the deregulation of the financial sector, and the decentralization of governance to weaken popular and national sovereignty (Brown, 2007). However, we can also understand neoliberalism as a global rationality aiming to transform institutions, social relations, and subjectivity. Following Michel Foucault's work on governmentality (Foucault, 2008), some authors claim that

> this norm enjoins everyone to live in a world of generalized competition ... it aligns social relations with the model of the market [and] it even transforms the individual, now called on to conceive and conduct him- or herself as an enterprise. (Dardot & Laval, 2014, p. 8)

---

[1] In May 2020, Sidewalk Labs cancelled its plan to build a smart neighbourhood in Toronto for economic uncertainty and lack of social acceptability. "The project, dubbed Quayside, still didn't have all of the government approvals it needed to go ahead. Toronto citizens and civic leaders had raised concerns about the privacy implications of the project and how much of the city's developing waterfront Sidewalk Labs wanted to control" (Carter & Rieti, 2020).

In this chapter, we will use the term "neoliberalism" in an all-encompassing way, as an ideology, a set of public policies, and a global rationality that contributes to the erosion of democratic language in favour of an "economization" of several aspects of life and social relations (Brown, 2015). In the urban realm, neoliberal rationality takes the form of an urban development model pushed by businesses, markets, innovation, and entrepreneurship. The neoliberal smart city is then "a high-tech variation of urban entrepreneurialism" (Hollands, 2008, p. 305). For example, the "hackathons" (prototyping events), which are closely linked to smart cities and the start-up economy, foster "the desire and belief in entrepreneurial life and technocratic governance, and also engender precarious, ambiguous and uncertain future for participants and prototypes" (Perng et al., 2017, p. 1). However, if some hackathons are motivated or used for commercial purposes, and though embedded in a neoliberal logic, they can also be used for civic purposes, such as the Code for America founded in 2009 and the Civic Hall New York, which focus on advancing civic tech and problem-solving for the public good.

Furthermore, techno-centric smart cities and neoliberalism are related through two interconnected phenomena: quantification of performance and austerity urbanism (Peck, 2015). First, the increasing competition between cities driven by globalization goes hand-in-hand with the rise of ranking systems and competitive tables that constrain cities to measure their global performance and "city score." The smart cities wheel system of 62 indicators is one example among many (Cohen, 2014). This imperative to render cities accountable, competitive, and measurable inevitably drives the tendency towards quantification. Hence,

> it is nevertheless rather obvious that the ranking-of-everything mentality upon which it rests is only possible in a city capable of collecting, analysing, and processing vast amounts of data. Thus, willingly or not, the smart city agenda along with the infrastructure of sensors and connectivity it promotes also opens the door to the kind of audit-obsessed quantification celebrated by neoliberalism. (Morozov & Bria, 2018, p. 10)

Second, when cities are constrained by austerity measures from national governments, increasing debt, or financial difficulties, they are often tempted to sub-contract public services or privatize infrastructures. At the same time, the rise of digital platforms like Uber and Airbnb accelerate the concentration of big data into the hands of private corporations, who enjoy a huge amount of knowledge on urban systems and become good candidates for running city services. Many forms of privatization undertaken by austerity urbanism are underway. For example, smaller cities from Florida to New Jersey are now paying Uber to offer subsidized rides to its inhabitants, while Washington, DC, already employs Uber to transport the disabled – deemed preferable (and cheaper) than investing in new bus lanes, trains, or other forms of public transportation. (Morozov & Bria, 2018, p. 16)

Furthermore, smart cities become increasingly dependent on corporations like Google as they subcontract their services, privatize their infrastructures, and give their data "for free" in exchange of expertise, algorithms, and management systems based on artificial intelligence technologies. "Such AI-powered services can then be used to further optimize how the city runs and operates – the city itself becomes a problem to finally be solved" (Morozov & Bria, 2018, p. 18).

In terms of participation, techno-centric neoliberal smart cities usually rely on a managerial view of participatory governance (Bacqué et al., 2005). In this model, urban governance is decentralized and involves numerous types of non-state actors, but economic actors usually have more influence on decision-making. Participation rarely invokes ideals of democracy or social justice and commonly invites several stakeholders to "collaborate" to solve city problems.

Many smart cities emphasize local participation and empowerment, but the underlying vision is often individualistic. Terms like "empowerment" can be given a neoliberal interpretation through promotion of individual capacity building, entrepreneurship, and access to opportunities in the market economy (Bacqué & Biewener, 2013). Even though techno-centric neoliberal smart cities often use the language of "people-centric," "user-centric," or "citizen-centric" development, we must analyse those catchphrases critically. Emphasis on the citizen, civil society, and local communities can be used like the "Big Society" rhetoric to justify dismantling the welfare state and transferring social responsibilities to individuals and enterprises (Williams et al., 2014).

To evaluate more precisely the quality of participation in smart cities, we will use the model of "smart citizen participation," which was elaborated to "measure smart citizen inclusion, participation, and empowerment in smart city initiatives in Dublin, Ireland" (Cardullo & Kitchin, 2019, p. 1). Following Arnstein's famous "participation ladder," which distinguishes different forms of participation that range from manipulation and consultation to partnership and citizen control (Arnstein, 1969), these authors claim that techno-centric and neoliberal discourses promote weak forms of participation like consumerism and non-participation. Of course, not all smart city initiatives in Dublin can be identified as technocratic or neoliberal, but many examples of digital participation can be categorized this way. Most "smart city initiatives are rooted in stewardship, civic paternalism, and a neoliberal conception of citizenship that prioritizes consumption choice and individual autonomy within a framework of state and corporate defined constraints that prioritize market-led solutions to urban issues" (Cardullo & Kitchin, 2019, p. 1). As a consequence, the citizen is often conceived as a user, data provider, patient, product, or consumer, and rarely as a genuine co-creator or decision-maker (see Table 1.1).

Fortunately, the techno-centric neoliberal model is not the only one. Digital technologies, participation, and environmental management can be articulated in several ways, so that the focus on a "technological push" can be replaced by a conception aiming to find a different balance between the various dimensions of city life. If the neoliberal smart city model "is more oriented towards growth, specifically economic growth driven by technological innovation" (Freeman, 2017, p. 60), others argue that "the sustainable city is focused more on finding equilibrium between its dimensions, or taking a triple bottom line perspective between people, planet and prosperity" (p. 17).

Promises of the Collaborative Sustainable City

The second model of smart cities is rooted in the framework of "sustainable cities." The term "sustainable development" first appeared in the 1987 report by the World Commission on the Environment and Development, which suggested a form of "development that meets the needs of the present without compromising the ability of future generations to meet their own needs" (WCED, 1987, p. 16). Since then, sustainability became a widespread notion used to integrate environmental concerns in spheres that ranged from science, education, and technology to urban development and public policies. More than a simple buzzword, it quickly became "a paradigm for thinking about the future in which environmental, societal, economic and cultural considerations are balanced in the pursuit of an improved quality of life" (Basera, 2016, p. 37772).

Sustainable development strategies can be developed at several scales, but they often take the form of local initiatives. Since the Rio Earth Summit in 1992, the Local Agenda 21 became a well-known formula insisting on the need for cities to identify priorities, elaborate action plans across different sectors, and specify goals for monitoring and reporting. Once again, the promotion of indicators to assess urban sustainability can justify an intensive use of sensors, data collection, algorithms, and hence smart city initiatives (Marsal-Llacuna et al., 2015). For example, the Aalborg Charter adopted in 1994, also known as the Charter of

Table 1.1 Scaffold of smart citizen participation

Forms and level of participation		Role	Citizen involvement	Political discourse/ framing	Modality	Dublin examples
Consumerism	Choice	Resident, consumer, product	Browse, consume, act	Capitalism, market, neoliberalism	Top-down, civic paternalism, stewardship, bound-to-succeed	Smart building/ district/ tech, personal data generated by tech
Non-participation	Therapy, manipulation	Patient, learner, user, data-point	Steered, nudged, controlled	Stewardship, technocracy, paternalism		Smart Dublin, Dublin bikes, traffic control

Source: Cardullo and Kitchin (2019, p. 5).

European Sustainable Cities and Towns Towards Sustainability, defines sustainability as a

> creative, local, balance-seeking process extending into all areas of local decision-making. It provides ongoing feedback in the management of the town or city on which activities are driving the urban ecosystem towards balance and which are driving it away. By building the management of a city around the information collected through such a process, the city is understood to work as an organic whole and the effects of all significant activities are made manifest. (Aalborg, 1994, p. 2)

From the perspective of sustainable cities, digital technologies represent only a means to achieve the higher goal of "urban sustainability." This notion was best defined by the ten Melbourne principles adopted during the 2002 Earth Summit in Johannesburg (UNEP, 2002):

1 Provide a long-term vision for cities based on sustainability; intergenerational, social, economic, and political equity; and their individuality.
2 Achieve long-term economic and social security.
3 Recognize the intrinsic value of biodiversity and natural ecosystems, and protect and restore them.

4 Enable communities to minimize their ecological footprint.
5 Build on the characteristics of ecosystems in the development and nurturing of healthy and sustainable cities.
6 Recognize and build on the distinctive characteristics of cities, including their human and cultural values, history, and natural systems.
7 Empower people and foster participation.
8 Expand and enable cooperative networks to work towards a common, sustainable future.
9 Promote sustainable production and consumption, through appropriate use of environmentally sound technologies, and effect demand management.
10 Enable continual improvement, based on accountability, transparency, and good governance.

A quick look at these principles shows that sustainable cities do not focus solely on ICT, innovation, business-led development, and economic growth, but also promote a global vision based on multidimensional equity, long-term economic and social security, the intrinsic value of nature, recognition of future generations, etc. This means that the sustainability discourse can be easily distinguished from the techno-centric and neoliberal model in terms of values. However, a critical approach needs to analyse more deeply the meaning of statements like "Empower people and foster participation," or fuzzy notions like "good governance" that can be used as "empty signifiers" (Offe, 2009).

Sustainable cities favour a type of urban development that is not led mainly by businesses and market forces, but by the modernization of public institutions. As with the model of "participatory modernization," these initiatives aim to improve the transparency of public administration, access to information, public servants' responsiveness, and inclusion of citizens in public management (Bacqué et al., 2005). One key idea is to "open" the smart city for the public good. For example, the organization Open North published an "Open Smart City Guide 1.0" developed through a collaborative research project with representatives, officials, and experts from Edmonton, Guelph, Montreal, and Ottawa (Lauriault et al., 2018). If the initiatives are more often top-down, rather than bottom-up, they try to change the relationship between people and local administrations in order to implement sustainable policies. From this perspective,

> a smart city is thus one that utilizes e-government, publishes open data and fosters an open data economy, creates citizen-centric dashboards about city performance, encourages citizen participation in reporting issues and

planning, enables urban test-bedding (wherein companies can trial new technologies aimed at improving urban services), actively nurtures start-up companies and accelerator programmes, promotes the use of ICT in education programmes, and actively leverages the technologies and data … to create new synergies, especially cross-sectoral approaches that break down departmental silos. (Kitchin, 2016, pp. 11–12)

The "cross-sectoral approach" of sustainable development implies a decentralized model of governance based on a "collaboration from diverse stakeholders to take a comprehensive approach to solving cities' complex challenges" (Freeman, 2017, p. 21). This collaborative model favours public-private-people-partnerships, aiming to have government, private sector, academia, and civil society actors collaborate with joint actions, programs, and projects (Lubell, 2015). Even though public institutions continue to play a key role in this model, they are called to work closely with businesses and ICT industries, universities and communities to stimulate innovation, implement robust policies, and sustain sustainable initiatives.

As many authors argued, collaborative urban governance is a central strategy to build sustainable cities (Kordas et al., 2015). Moreover, there seems to be a convergence of sustainable and smart city models, because collaborative sustainability governance is strongly supported by ICT (Termeer & Bruinsma, 2016). It is no coincidence that the "smart-sustainable city" concept recently appeared in the literature, to define

an innovative city that uses information and information and communication technology (ICT) and other means to improve quality of life, efficiency of urban operations and services, and competitiveness, while ensuring that it meets the needs of present and future generations with respect to economic, social, environmental as well as cultural aspects. (ITU, 2016, p. 52)

In terms of participation, the model of collaborative sustainable cities puts greater emphasis on citizen engagement than techno-centric neoliberal cities do. For example, the city of Grenoble, France, which is a candidate for the European Green Capital Award in 2022, has several mechanisms of public participation like independent citizen councils, participatory budgeting, a local inhabitants fund, and a citizens initiative that can force a public vote in city council (Ville de Grenoble, 2017). In Canada, Montreal won the $50 million prize of the Canadian Smart City challenge in May 2019 to work on a community process related to mobility and food security – which are both sustainability issues.

In this model, empowerment is not only seen through the lens of entrepreneurship and market economy, but is also conceived from the social liberal perspective, which emphasizes individual autonomy, social cohesion, equal opportunities, struggle against poverty, and good governance (Bacqué & Biewener, 2013). Citizens have more voice in the participatory processes, through online tools and smart city initiatives. The citizen is seen not only as a consumer or passive user, but also as a participant, tester, and proposer. For example, Fix-Your-Street allows citizens to "report the location of issues that need to be addressed (such as potholes, graffiti, broken streetlights, illegal dumping)," while Smart Dublin has "an advisory network of forty key stakeholders drawn from government, companies, universities and civil society that meets twice a year to offer constructive feedback on Dublin's smart city initiatives" (Cardullo & Kitchin, 2019, p. 8).

Nonetheless, Arnstein argues that mechanisms of information and consultation are limited to "tokenism" because they remain largely "top-down" and do not bring a real share of decision-making power between city administration and citizens. "When participation is restricted to these levels, there is no follow-through, no 'muscle,' hence no assurance of changing the status quo" (Arnstein, 1969, p. 217). In short, collaborative sustainable cities use digital technologies to improve transparency (e-government), "governments-to-citizens" relations (dashboards) and "citizens-to-governments" interactions (Fix-Your-Street), but they do not challenge the structural constraints of the dominant political and economic system.

That may not represent a problem in itself, but if we take the "principle of affected interests" seriously (Fung, 2013) and see how we can include citizens in democratic governance so that they can have a real influence on decisions that affect their interests, we need to challenge the traditional structure of representative democracy where citizen participation is separated from authoritative decisions in the political institutions (Durand Folco, 2017b). That doesn't mean that citizens must participate directly in all decisions that affect every aspect of their daily life, but it is necessary to go beyond weak forms of participation in order to reform institutions and build a participatory society (Pateman, 2012).

Instead, digital participation in collaboration cities takes the form of "social monitoring," which "entails an outsourcing of functions that are normally performed by experts and professions to the broad public, and soliciting back services, suggestions, solutions, observations, or ideas" (Fung et al. 2013, p. 42). Overall, collaborative governance implies middle-range forms of participation and tokenism, but rarely forms of partnership, delegated power, and citizen control (see Table 1.2).

Table 1.2 Scaffold of smart citizen participation

Forms and level of participation		Role	Citizen involvement	Political discourse/ framing	Modality	Dublin examples
Tokenism	Placation	Proposer	Suggest	Participation, co-creation	Top-down, civic paternalism, stewardship, bound-to-succeed	Fix-Your-Street, Dublin Advisory Network
	Consultation	Participant, tester	Feedback	Civic engagement		CIVIK, Smart Stadium
	Information	Recipient	Browse, consume, act			Dublinked, Dublin Dashboard

Source: Cardullo and Kitchin (2019, p. 5).

The Limits of Smart Sustainable Cities

Collaborative sustainable cities are more interesting than techno-centric neoliberal smart cities from the social and ecological point of view. Smart sustainable cities seem to integrate the best aspects of digital technologies and cross-sectoral initiatives that embrace city life. Although there can be synergies between the sustainable and the smart city concepts, there is a risk that "combining the two perspectives ... represents a further dilution of sustainable development goals" (Freeman, 2017, p. 60). Instead of focusing on urban sustainability as an overarching goal and conceiving ICT as one means among many to achieve it, "smart sustainable cities" may blur the lines between means and ends, and involuntarily introduce techno-centric and/or neoliberal elements into the collaborative sustainable cities discourse. In other words, the "collaborative city" could be midway on a continuum between the techno-centric smart city and the sustainable city.

However, this warning against the risk of "contamination" of sustainable development by technocracy and neoliberalism is not an argument against collaborative sustainable cities per se. To assess the relevance of collaborative sustainable cities, it is essential to formulate explicitly normative principles and analyse if this model can satisfy them. From the perspective of strong sustainability, democracy, and social justice, we claim that the collaborative sustainable cities discourse presents significant flaws, mostly because its emphasis on "equilibrium" obscures systemic conflicts and power relations between actors.

For example, even though sustainable cities seek an equilibrium between economic, social, and environmental dimensions of urban development, there still could be a contradiction between unlimited economic growth and ecological sustainability (Abraham et al., 2011). The perception of compatibility or incompatibility between those two aspects depends on whether actors adopt a strong or weak interpretation of sustainability. Strong sustainability means that "natural capital" (like ecosystems and non-renewable resources) cannot be replaced by "human capital" (technological development, labour, knowledge), while weak sustainability means that natural capital can be substituted with human capital (Pearce & Atkinson, 1993). In other words, weak sustainability usually implies some form of technological solutionism, because natural resources are seen a replaceable by innovation and digital technologies, while the strong sustainability perspective is profoundly sceptical about the possibilities of overcoming the limits of continuous economic growth (Meadows et al., 2004).

Overall, "reformist responses" to ecological challenges like administrative rationalism, liberal democracy, green economy, sustainable development, and ecological modernization embrace a weak form of sustainability (Dryzek & Schlosberg, 2005). Hence, weak sustainability is the dominant view in public debate in America, Europe, and even Asia (Barua & Khataniar, 2015).

> The eco-modernization version of a sustainable city predominant in Europe argues that green economic growth can be compatible with ecological integrity and social equity. This claim rests on the perceived ability of digital technologies to drive the dematerialization of the economy, by shifting from resource-intensive to knowledge-intensive forms of economic activity.
> (Freeman, 2017, p. 60)

However, digital infrastructures have a significant impact on the environment (Hodgson, 2015) because of the colossal energy consumption, increasing exploitation of rare-earth metals, and the Jevons paradox also known as the "rebound effect," which implies that efficiency gains in resource use will be offset by increasing overall consumption (Flipo & Gossart, 2009; Owen, 2012). If we take sustainability seriously and seek to transform cities in order to make them more resilient, it is necessary to avoid technological solutionism (Morozov, 2013) or unlimited growth, and to initiate a socio-ecological transition to a new economic model (Wright, 2010).

From a democratic perspective, the framework of collaborative governance aims to include several stakeholders to resolve complex problems, but it doesn't specify which actors should have the final word or make decisions. While the government paradigm supposed that representatives should have the authority to make publicly binding decisions, the shift to governance implies more flexible, networked decision-making processes where authority is devolved to non-state actors and task-specific institutions at different scales (Bellamy & Pallumbo, 2010). In this context, should the government, businesses, or citizens have the greatest weight in urban development? Even if collaborative sustainable cities promote citizen empowerment, do they have real decision-making power? If we take the example of living labs in which users are supposed to be at the centre of innovation, the living labs can be in fact centred on the interests of firms, public authorities, or citizens (Arnkil et al., 2010), depending on who will get the major benefits of economic resources, legitimacy, recognition, rewards, etc.

Thus, the will to include the private sector, public institutions, and civil society in governance does not mention the power relations are, and indeed the structures of domination between different organizations and actors. In a context of rising economic inequalities, unequal access to decision-making spaces, neoliberal urbanism (Peck et al., 2009), and the rise of digital capitalism (De Grosbois, 2017), there are plenty of struggles over control of ICT and city development. The collaborative "win-win" strategy and conciliating approach of sustainable cities is not framed to identify social inequalities, systemic causes of oppressions, and power relationships in general. Hence, the discourse of collaborative sustainable cities makes urban development vulnerable to the influence of economic and political elites, and could indeed reproduce injustices.

The same problem arises with the tensions between competitiveness and social cohesion. Collaborative sustainable cities take into account the social dimension of urban development, but this model still keeps a neoliberal framework based on competitiveness and attractiveness for private investments. For example, "the current economic disparity in the city has incentivized the City of Malmö to shift its efforts from more environmentally focused projects to socially oriented ones, all while maintaining its commitment to building an attractive city for entrepreneurs and businesses" (Freeman, 2017, p. 53, referring to Nylund, 2014). We can see the same example in Montreal, where the green and social measures put forward by the administration of Mayor Valérie Plante since 2017 are linked to an economic development strategy based mostly on entrepreneurship, international competitiveness, technological innovation, and artificial intelligence (Ville de Montréal, 2018).

The underlying political vision of sustainable development can be related to social liberalism, otherwise called the "Third Way" advertised by Bill Clinton, Tony Blair, and Anthony Giddens. This hybrid perspective, aiming to go beyond the Left and Right divide (Giddens, 1998) through the promotion of market economy, social capital, collaboration, and the empowerment of local communities (Bacqué & Biewener, 2013), cannot acknowledge systemic domination and propose a real challenge to the hegemony of neoliberalism. As Susan Fainstein notes,

> The assumption underlying the Third Way is that increasing cohesion – which seems to mean creating a more diverse, tolerant and equitable urban society – will result in economic success as well. The recipe appears to be decentralized governance through partnerships among all sectors of the population. Even while accepting competition as the context of urban development, it is a denial of lines of conflict and exercises of power. (Fainstein, 2001, pp. 887–888)

In short, the collaborative sustainable cities model does not adequately address many important ecological, democratic, and social concerns. It aims to reconcile economic, technological, social, and environmental dimensions in a balanced world view where all private, public, and civil society actors should work hand-in-hand, thus discarding antagonisms, social conflicts, and contradictions between opposite imperatives. That doesn't mean that actors from different backgrounds should not collaborate at all in order to solve complex social, technical, economic, and political problems, but we reject the idea that all stakeholders usually participate on an equal footing, and could easily build a common view on the form of the sustainable city. In this sense, we do not suggest that all dimensions of collaborative sustainable cities should be thrown away, but that it represents an insufficient model if we embrace the principles of strong sustainability, radical democracy, and social justice, which will necessarily imply trade-offs between conflicted interests and challenge patterns of domination and privileges. Now that we have identified some weaknesses of this paradigm, we will present a third model that could become a real alternative model to techno-centric neoliberal smart cities.

The Rise of Citizen-Centred Rebel Cities

A new citizen-centred model is emerging from a political context in which social movements recently won local elections with the hope to bring a "radical change" in municipal administration. After the global financial crisis of 2007–2008, a wave of popular mobilizations (15-M movement in Spain, or the Occupy movement) suddenly appeared to denounce political corruption, wealth inequalities, and the hegemony of the 1 per cent. Slogans like ¡Democracia Real YA!, public space occupations, online self-organizing, and direct democracy in citizen assemblies challenged the political and economic status quo. In 2015, citizen platforms stemming from social movements, radical left-wing parties, and local associations won municipal elections in major cities in Spain, like Barcelona, Madrid, Zaragoza, Cádiz, etc.

These cities were quickly named "rebel cities" in public debate (Lamant, 2016), because a new generation of activists entered city halls to challenge neoliberal urbanism, traditional representative democracy, conservative national government, and capitalist market economy. Rebel cities aim to defend social rights, democratize the economy, and defend the commons against the assaults of economic elites and the far right. Hence, they represent a third and distinct model, which is explicitly opposed to the techno-centric neoliberal smart city perspective and represents a much more radical project than the collaborative sustainable city paradigm. This alternative model is related to municipalism, which is a political movement that considers the municipality, commune, or local government as a lever for the democratic transformation of social, economic, and political life (Durand Folco, 2017a).

Before going further, we need to distinguish between the citizen-centric rebel city, like Barcelona and Madrid, which represents the "actually existing municipalism," and the large, emerging, and international political movement that is organizing itself through transnational networks. The first International Municipalist Summit, "Fearless Cities," held in June 2017 in Barcelona, indicated that municipalism is not a marginal movement confined to Spain, but an emerging global network of local initiatives aiming to "take back" the city and experiment with new forms of radical democracy (Baird et al., 2019). If the "Rojava revolution" in Syrian Kurdistan is probably the best example of municipalist experiment operating on a large scale in the world right now (Knapp et al., 2016), many other cities saw the rise of "radical civic platforms standing up to entrenched political interests" like Valparaíso (Chile), Beirut (Lebanon), Zagreb (Croatia), etc. (Reyes, 2017).[2] In North America, the best example of a "rebel city" is Jackson (Mississippi), which has been governed by a socialist mayor since 2014, in relation to popular citizen assemblies and a global strategy aiming for self-determination and economic democracy for Black people (Akuno & Nangwaya, 2017).

---

2 For a better overview of the municipalist movement around the world, see the Minim database, launched in October 2019 to promote municipalism through sharing practical and theoretical knowledge with the support of a community of activists, scholars, journalists, and public officials: https://minim-municipalism.org/.

Since the municipalist movement is new and mainly absent from North America at the institutional level, it is more difficult to find municipalities that could be given the "rebel cities" label in Canada. However, if no city government has been seized by radical citizen platforms yet, there is a growing libertarian municipalist movement organizing through the work of activists in Canada, the United States, and Mexico (Durand Folco & Van Outryve, 2019). Furthermore, there is growing interest in municipalism and the "urban commons" in Quebec. For example, CITES, which supports the international gathering, sharing, and transfer of knowledge and best practices in the social economy, recently published a comparative study on the urban commons in Barcelona and Montreal (Durand Folco et al., 2019). In other words, the rebel cities "spirit" is slowly influencing activists, researchers, individual officials, and representatives in Montreal, even though this perspective is not yet institutionalized in a common framework.

While neoliberal smart cities focus on technological innovation and economic growth, and collaborative sustainable cities seek to find a balance between economic, social, and environmental dimensions of urban development, the model of rebel cities is centred on the collective needs of citizens and social concerns (Cliché, 2017). Digital technologies are not seen mainly as a way to solve every city problem (as with technological solutionism), or a way to modernize public institutions, but as a means to foster citizen power, popular mobilization, democratic governance, radical transparency, etc. As Kitchin notes, this alternative

> conception of a smart city is one that uses digital technologies and ICT to promote a citizen-centric model of urban development and management that promotes social innovation and social justice, civic engagement and hacktivism, and transparent and accountable governance ... Here, there is an emphasis on fostering civic hacking and hackathons; participatory planning and community development; open source platforms, software and data; freedom of information; crowd sourcing and communal action; and digital and data literacy. (Kitchin, 2016, p. 12)

One key principle beginning to be used is "technological sovereignty," which refers to "citizens' capacity to have a say and participate in how the technological infrastructure around them operates and what ends it serves" (Morozov & Bria, 2018, p. 22). Technological sovereignty can be applied through various forms, but it usually challenges the hegemony of big corporations and collaborative economy platforms, like Barcelona did, when it "fined Airbnb €600,000 for continuing to advertise unlicensed flats on its platform" in 2016 (Burgen, 2017).

Another way to concretize technological sovereignty is through re-municipalization of digital infrastructures and public services (Kishimoto & Petitjean, 2017). Even though the re-municipalization of digital infrastructures is not as easy to implement as the re-municipalization of public transport, waste management, or water systems, cities can use free software and open-source alternatives. For example, Barcelona decided to switch to Linux, Libre Office, and Open Xchange (Leonard, 2018). Furthermore, it is essential to transform the data ownership regimes in order to foster "city data commons" and to build alternative models to capitalist platforms like Uber and Airbnb, like the "platform cooperativism" movement, which advocates for worker-owned and commons-based digital platforms (Scholz & Schneider, 2016). Sharing Cities Seoul Initiative and FairBnB in Amsterdam are probably the most famous examples of projects aiming to build smart cities on social and a solidarity economy scheme.

This objective to democratize the economy is closely related to the defence of social rights. This can take the form of free public transit (Deillhem & Prince, 2018), massive funding of social and affordable housing, as in Red Vienna (Gruber, 1991), complimentary local currencies (Blanc, 2011), like the Toronto Dollar, which existed from 1998 to 2013 (Herpel, 2008), and the local currency called "Blé" in Québec City, which was launched in 2018 (Lê-Huu, 2018). There are even municipal basic income projects, as in Oakland (California), Utrecht, Livorno, and Glasgow (Morozov & Bria, 2018). Such measures do not necessarily imply digital technologies, but they do reflect a more general sensibility to social justice by citizen-centric rebel cities. For example, rebel cities can also disobey and refuse measures taken by different levels of government, like the "sanctuary cities" in the United States who refused to cooperate with Trump's anti-immigration decisions (Dinan, 2018). In the same spirit, local governments in Europe decided to become "free zones" against free trade agreements following the "Barcelona declaration" signed by 60 cities (Chapelle, 2017).

> By declaring their cities and counties CETA- and TTIP-free zones, mayors and local politicians affirm their will ... to make citizens' voices heard by their governments and by the European Commission, defending local communities and democratic institutions as spaces for debate and decision-making. (FoEE, 2016)

From the sustainability point of view, citizen-centric rebel cities go beyond the sustainable development discourse, as many activists, ecologists, and other citizen movements often share a "green social critic" of the economic system (Dryzek & Schlosberg, 2005), as with the slogan

System Change, Not Climate Change. Hence, many discourses like bio-regionalism (Magnaghi, 2005, 2014), eco-socialism (Kovel, 2007), environmental justice (Schlosberg, 2007), degrowth (D'Alisa et al., 2014), and social ecology (Bookchin, 1982), are often linked to municipalism and rebel cities initiatives (Durand Folco, 2017a). Sustainability is then usually conceived in its strong version, and the ecological transition initiatives – like transition towns (Hopkins, 2011) – will favour a global degrowth of energy consumption, re-localization of the economy, grassroots innovations, and overall democratization of society. "Thus, other versions of a sustainable city have been advanced with the idea of creating environmental sustainability through a stable state economy or even a degrowth economy" (Freeman, 2017, p. 60).

On the participation dimension, citizen-centric rebel cities will have a more radical conception of empowerment, associated with self-determination, wealth redistribution, individual and collective emancipation, social justice, grassroots organization, and social transformation (Bacqué & Biewener, 2013). The model of participatory democracy will then be related to a strong presence of social movements, more radical and direct forms of democratic procedures (citizen assemblies, direct digital democracy, participatory budgeting), an "inversion of social priorities" favouring the interest of vulnerable groups and popular classes, and a politicized perspective of urban development (Bacqué et al., 2005).

Although the rebel city model includes elements of collaborative governance like social monitoring or online platforms to collect citizens' data, suggestions, and preferences, this approach aims to build a more radical form of digital democracy (Durand Folco, 2016). One good example is Decide Madrid, an open-source direct democracy platform launched by the City of Madrid in 2015. It enables

> citizens to propose, debate, prioritize, and implement city policy ... as a bottom-up proposal mechanism in which anyone can propose and debate, but only residents of Madrid have a binding vote. A deliberative forum is integrated into the citizen proposal mechanism, along with, more recently, a participatory budgeting tool, with the city investing €10 million in citizens' proposals thus far. (Morozov & Bria, 2018 p. 51)

To use Arnstein's terminology, the citizen-centric cities model promotes stronger forms of partnerships, co-creation, delegated power, and citizen control. Once again, the smart citizen participation model built by Cardullo and Kitchin to evaluation digital initiatives in Dublin can help us to understand the vision of citizen power beyond techno-centric and collaborative cities (see Table 1.3).

Table 1.3 Scaffold of smart citizen participation

Forms and level of participation		Role	Citizen involvement	Political discourse/ framing	Modality	Dublin examples
Citizen power	Citizen control	Leader/ member	Ideas, vision, leadership, ownership, create	Rights, social/ political citizenship, deliberative democracy, commons	Initiative, bottom-up, collective, autonomy, experimental	Code for Ireland, Tog Civic Hacking, Hackathons, Living Labs, Dublin Beta
	Delegated power	Decision-maker, maker				
	Partnership	Co-creator	Negotiate, produce			

Source: Cardullo and Kitchin (2019, p. 5).

Following the actions undertaken by rebel cities to counteract the techno-centric neoliberal smart city agenda, Evgeny Morozov and Francesca Bria (the Chief Technology and Digital Innovation Officer for the City of Barcelona) propose nine principles to promote a model based on public control, democratic governance, and citizens' self-organization. Just like the ten Melbourne Principles, which established the baseline of sustainable cities, we can consider those principles as the model of a citizen-centred rebel cities model (Morozov & Bria, 2018). If it isn't possible to analyse each principle in detail, this list offers a global perspective that summarizes the significant aspects of the rebel cities political project.

The principles of the right to the digital city are as follows:

1 Promote alternative data ownership regimes.
2 Move information service to open source, open standards, and agile delivery.
3 Transform procurement to make it ethical, sustainable, and innovative.
4 Control digital platforms.
5 Build and grow alternative digital infrastructures.
6 Develop cooperative models of service provision.
7 Maximize innovation with public value.
8 Rethink local welfare schemes and complementary currency systems.
9 Promote digital democracy and digital sovereignty. (Morozov & Bria, 2018, p. 30)

As you probably noticed, the authors' failure to mention any sustainability, ecological transition, or environmental/climate justice principles within their framework constitutes an important missing point or dead angle of this model. The citizen-centric model places emphasis on the collective control of digital technologies, therefore downplaying the environmental dimension of the city. However, we could easily add a tenth principle in order to expand this perspective towards a "right to the digital, ecological, and democratic city."

Conclusion

Obviously, this threefold typology of smart, sustainable, and rebel cities can seem a bit oversimplified. Smart cities are not always techno-centric and neoliberal. Sustainable cities do not systematically refer to weak forms of sustainability and collaborative governance. And rebel cities are not in themselves the great response to all the social, ecological, democratic, and economic challenges of the twenty-first century.

Nonetheless, this ideal-type approach aims to distinguish three broad discourses, perspectives, or models that can be more or less applied in practice by different actors. This analytical framework, which is not devoid of normative concerns for social justice, democracy, and sustainability, aims to make explicit the competing visions of the city's future. As unequal and conflicting social, economic, and political actors actually try to orient urban development, it is theoretically useful and strategically important to make a more explicit broad set of presuppositions, ideals, and interests that could lead us in diverging directions. Finally, it would also be possible to conceive a continuum between two opposite views of the city, collaborative sustainability being somewhere in the middle between the techno-centric neoliberal smart cities and the citizen-centric rebel city, each model having its own internal logic and specific dimensions (see Table 1.4).

In short, we tried to amplify the differences between techno-centric neoliberal smart cities, collaborative sustainable cities, and citizen-centric rebel cities in order to lighten the potentials and pitfalls of our urban futures. This chapter did not seek to provide a prognosis on which model will triumph in the next years, nor to sketch a blueprint of the ideal city. The main goal was to provide a critical analysis to render more explicit the political views embedded in digital technological discourses. Overall, the implicit ideologies implied in the various smart cities proposals try to define who should direct urban development. "Who decides? Who controls the money, the design of projects, how services are run?" (Wilcox, 1994, p. 3).

In times of great dangers with rising authoritarian populisms, enormous social inequalities, surveillance capitalism, and perspectives of ecological and civilizational collapse (Diamond, 2005; Servigne & Stevens,

Table 1.4 Three models of the city

	Techno-centric neoliberal city	Collaborative sustainable city	Citizen-centric rebel city
Central dimension	Technological	Balance between economic, social, and environment	Social
Main concerns and issues	Innovation, entrepreneurship, efficiency, city services optimization	Resource management, inclusion, modernization of public institutions	Social justice and rights, democratization, technological sovereignty
Model of governance	Managerial governance, data-driven governance	Collaborative governance	Democratic governance, radical democracy
Level of participation	Non-participation, consumerism	Information, consultation, placation	Partnership, delegated power, citizen control
Role of the citizen	User, data-point, resident, consumer	Participant, tester, proposer	Co-creator, decision-maker, leader
Model of empowerment	Neoliberal empowerment	Social liberal empowerment	Radical empowerment
Environmental discourse	Ecological modernization, techno solutionism	Sustainable development	Social ecology, environmental justice
Type of sustainability	Weak	Weak or strong	Weak or strong
Use of digital technologies	Solving city problems, optimization of everything	Modernization, transparency	Popular mobilization, emancipation
Main actors	ICT businesses, start-ups, digital platforms	City administration, public-private-people partnerships	Citizens, social movements
City examples	Masdar, Songdo	Amsterdam, Malmö	Barcelona, Madrid

2015; Wallace-Wells, 2019), the role, instrumentalizations, promises, and illusions of digital solutions should not be underestimated. As Nancy Fraser notes, the primary goal of critical theory is to foster "the self-clarification of the struggles and wishes of the age" (Fraser, 1985, p. 97). For better or for worse, smart cities are now at the front lines of the digital, social, and urban battlefield.

REFERENCES

Aalborg Charter. (1994). *Charter of European cities & towns towards sustainability.* http://www.sustainablecities.eu/the-aalborg-charter/

Abraham, Y.-M., Marion, L., Philippe, & H. (Eds.). (2011). *Décroissance versus développement durable. Débats pour la suite du monde.* Écosociété.

Akhtar, P., Tse, M., Khan, Z., & Rao-Nicholson, R. (2016). Data-driven and adaptive leadership contributing to sustainability of global agri-food supply chains connected with emerging markets. *International Journal of Production Economics, 181*, 392–401. https://doi.org/10.1016/j.ijpe.2015.11.013

Akuno, K., & Nangwaya, A. (2017). *Jackson rising: The struggle for economic democracy and Black self-determination in Jackson, Mississippi.* Daraja Press.

Albito, V., Berardi, U., & Dangelico, R. M. (2015). Smart cities: Definitions, dimensions, performance, and initiatives. *Journal of Urban Technology, 22*(1), 3–21. https://doi.org/10.1080/10630732.2014.942092

Amin, A., & Thrift, N. (2002). *Cities: Reimagining the urban.* Polity Press.

Arnkil, R., Järvensivu, A., Koski, P., & Piirainen, T. (2010). Exploring quadruple helix: Outlining user-oriented innovation models. Institute of Social Research, University of Tampere.

Arnstein, S. (1969). A ladder of citizen participation. *Journal of the American Institute of Planners, 35*(4), 216–224. https://doi.org/10.1080/01944366908977225

Audet, R. (2016). Discours autour de la transition écologique. In M.-J. Fortin, Y. Fornis, & F. L'Italien (Eds.), *La transition énergétique en chantier* (pp. 11–30). Les configurations institutionnelles et territoriales de l'énergie, Presses de l'Université Laval.

Bacqué, M.-H., & Biewener, C. (2013). *Empowerment. Une pratique émancipatrice?* La Découverte.

Bacqué, M.-H., Rey, H., & Sintimer, Y. (Eds.). (2005). *Gestion de proximité et démocratie participative. Une perspective comparative.* La Découverte.

Baird, K. S., Junque, M., & Barcelona en comu. (2019). *Fearless cities: A guide to the global municipalist movement.* New Internationalist.

Barua, A., & Khataniar, B. (2015). Strong or weak sustainability: A case study of emerging Asia. *Asia-Pacific Development Journal, 22*(1), 1–31. http://doi.org/10.18356/9b582978-en

Basera, N. (2016). Sustainable development: A paradigm shift with a vision for the future. *International Journal of Current Research, 8*(9), 37772–37777.

Bellamy, R., & Pallumbo, A. (Eds.). (2010). *From government to governance.* Routledge.

Bibri, S. E. (2018). *Smart sustainable cities of the future: The untapped potential of big data analytics and context-aware computing for advancing sustainability.* Springer International Publishing.

Blanc, J. (2011). Classifying "CCs": Community, complementary and local currencies' types and generations. *International Journal of Community Currency Research, 15*, 4–10. http://doi.org/10.15133/j.ijccr.2011.013

Bookchin, M. (1982). *The ecology of freedom: The emergence and dissolution of hierarchy*. Cheshire Books.

Brown, W. (2007). *Les habits neufs de la politique mondiale: Néolibéralisme et néo-conservatisme*. Les Prairies ordinaires.

Brown, W. (2015). *Undoing the demos: Neoliberalism's stealth revolution*. Zone Books.

Burgen, S. (2017, 2 June). Barcelona cracks down on Airbnb rentals with illegal apartment squads. *The Guardian*.

Cardullo, P., & Kitchin, R. (2019). Being a "citizen" in the smart city: Up and down the scaffold of smart citizen participation in Dublin, Ireland. *GeoJournal, 84*, 1–13.

Carter, A., & Rieti, J. (2020, 7 May). Sidewalk Labs cancels plan to build high-tech neighbourhood in Toronto amid COVID-19. CBC News. https://www.cbc.ca/news/canada/toronto/sidewalk-labs-cancels-project-1.5559370

Chapelle, S. (2017, 7 March). Ces zones libres en Europe qui privilégient les alternatives locales face au néolibéralisme global. Bastamag. https://www.bastamag.net/Ces-zones-libres-en-Europe-qui-privilegient-les-alternatives-locales-face-au

Cliche, D. (2017). La ville intelligente au service du bien commun. Lignes directrices pour allier l'éthique au numérique dans les municipalités au Québec. Commission de l'éthique en science et en technologie, Gouvernement du Québec.

Cohen, B. (2014, 20 November). The smartest cities in the world 2015: Methodology. Fast Company. https://www.fastcompany.com/3038818/the-smartest-cities-in-the-world-2015-methodology

Couch, C. (2009). Privatised Keynesianism: An unacknowledged policy regime. *The British Journal of Politics and International Relations, 11*(3), 382–399. https://doi.org/10.1111/j.1467-856X.2009.00377.x

D'Alisa, G., Demaria, F., & Kallis, G. (Eds.). (2014). *Degrowth: A vocabulary for a new era*. Routledge.

Dardot, P., & Laval, C. (2014). *The new way of the world: Neoliberal society*. Verso.

De Grosbois, P. (2017). *Les batailles d'Internet. Assauts et résistance à l'ère du capitalisme numérique*. Écosociété.

Deillhem, J., & Prince, J. (Eds.). (2018). *Free public transit: And why we don't pay to ride escalators*. Black Rose Books.

de Jong, W. M., Joss, S., Shraven, D., Changjie, Z., & Weijnen, M. (2015). Sustainable – smart – resilient – low carbon – eco – knowledge cities: Making sense of a multitude of concepts promoting sustainable urbanization. *Journal of Cleaner Production, 109*, 25–38. https://doi.org/10.1016/j.jclepro.2015.02.004

Diamond, J. (2005). *Collapse: How societies choose to fail or succeed*, Penguin Books.

Dinan, S. (2018, 10 May). Half of all Americans now live in "sanctuaries" protecting immigrants. *Washington Times*. https://www.washingtontimes.com/news/2018/may/10/half-of-americans-now-live-in-sanctuaries/

Durand Folco, J. (2016). Vers une démocratie numérique radicale. *Ithaque, 19*, 189–209. http://hdl.handle.net/1866/16171

Durand Folco, J. (2017a). *À nous la ville! Traité de municipalisme*, Écosociété.

Durand Folco, J. (2017b). *Transformer la ville par la démocratie participative. L'exemple des conseils de quartier décisionnels* [Thèse de doctorat]. Université Laval.

Durand Folco, J., L'Allier, M.-S., Audet, R., Guarner, L. E., Fouss, B., Furukawa Marques, D., Gauthier, É., Tadjine, N., Veillette, J., & Butzbach, C. (2019). Les communs urbains. Regards croisés sur Montréal et Barcelone, ouvrage collectif produit par le Centre international de transfert d'innovations et de connaissances en économie sociale et solidaire CITIES, Montréal. https://www.passerelles.quebec/publication/2019/les-communs-urbains-regards-croises-sur-montreal-et-barcelone/

Durand Folco, J., & Van Outryve, S. (2019, 14 October). Symbiosis: Federating municipalist movements in North America for real democracy. Minim. https://minim-municipalism.org/magazine/symbiosis-federating-municipalist-movements-in-north-america-for-real-democracy/

Dryzek, J. S., & Schlosberg, D. (Eds.). (2005). *Debating the earth: The environmental politics reader* (2nd ed.). Oxford University Press.

Fainstein, S. (2001). Competitiveness, cohesion, and governance: Their implications for social justice. *International Journal of Urban and Regional Research, 25*(4), 884–888. https://doi.org/10.1111/1468-2427.00349

Flipo, F., & Gossart, C. (2009). *Infrastructure numérique et environnement: l'impossible domestication de l'effet rebond, Terminal*. L'Harmattan.

Fraser, N. (1985). What's critical about critical theory? The case of Habermas and gender. *New German Critique, 35*, 97–131.

Fraser, N. (2017). From progressive neoliberalism to Trump – and beyond. *American Affairs, 1*(4), 46–64.

Foucault, M. (2008). *The birth of biopolitics: Lectures at the Collège de France (1978–1979)*. M. Senellart (Ed.). Palgrave Macmillan.

Freeman, G. (2017). *The origin and implementation of the smart-sustainable city concept: The case of Malmö, Sweden*. [MA thesis]. IIIEE Theses, The International Institute for Industrial Environmental Economics.

Friends of the Earth Europe (FoEE). (2016, 7 September). 2000 CETA and TIPP-free zones in Europe. http://foeeurope.org/2000-CETA-and-TTIP-free-zones-in-Europe

Fung, A. (2013). The principle of affected interests and inclusion in democratic Governance. In J. Nagel & R. Smith (Eds.), *Representation: Elections and beyond* (pp. 236–268). University of Pennsylvania Press.

Fung, A., Gilman, H. R., & Skhabatur, J. (2013a). Six models for the internet + politics. *International Studies Review, 15,* 30–47. https://doi.org/10.1111/misr.12028

Giddens, A. (1998). *Beyond left and right: The future of radical politics.* Polity Press.

Giffinger, R., Fertner, C., Kramar, H., Kalasek, R., Pichler-Milanovic, N., & Meijers, E. (2007). *Smart cities: Ranking of European medium-sized cities.* Center of Regional Science.

Grossi, G., & Pianezzi, D. (2017). Smart cities: Utopia or neoliberal ideology? *Cities, 69,* 79–85. http://doi.org/10.1016/j.cities.2017.07.012

Gruber, H. (1991). *Red Vienna: Experiment in working-class culture, 1919–1934.* Oxford University Press.

Haarstad, H. (2016). Constructing the sustainable city: Examining the role of sustainability in the "smart city" discourse. *Journal of Environmental Policy & Planning, 19*(4), 423–437. https://doi.org/10.1080/1523908X.2016.1245610

Hajer, M. A. (1995). *The politics of environmental discourse: Ecological modernization and the policy process.* Oxford University Press.

Herpel, M. (2008, 18 April). The Toronto dollar: Community alternative dollar. *California Chronicle.*

Hodgson, C. (2015, 13 November). Can the digital revolution be environmentally sustainable? *The Guardian.*

Höjer, M., & Wangel, J. (2015). Smart sustainable cities: Definition and challenges. In L. Hilty & B. Aebischer (Eds.), *ICT innovations for sustainability: Advances in intelligent systems and computing* (pp. 333–349). Springer.

Hollands, R. (2008). Will the real smart city please stand up? Intelligent, progressive or entrepreneurial? *City, 12*(3), 303–320. https://doi.org/10.1080/13604810802479126

Hopkins, R. (2011). *The transition companion: Making your community more resilient in uncertain times.* Chelsea Green Publishing.

Ibrahim, M., El-Zaart, A., & Adams, C. (2018). Smart sustainable cities roadmap: Readiness for transformation towards urban sustainability. *Sustainable Cities and Society, 37,* 530–540. https://doi.org/10.1016/j.scs.2017.10.008

International Telecommunication Union (ITU). (2016). *Shaping smarter and more sustainable cities: Striving for sustainable development goals.* ITU-T Focus Group on Smart Sustainable Cities.

Kishimoto, S., & Petitjean, O. (2017). *Reclaiming public services: How cities and citizens are turning back privatisations.* Transnational Institute. https://www.tni.org/files/publication-downloads/reclaiming_public_services.pdf

Kitchin, R. (2013). The real-time city? Big data and smart urbanism. *GeoJournal, 79*(1), 1–14. http://doi.org/10.1007/s10708-013-9516-8

Kitchin, R. (2016). *Getting smarter about smart cities: Improving data privacy and data security.* Data Protection Unit, Department of the Taoiseach.

Knapp, M., Ayboga, E., & Flach, A. (2016). *Revolution in Rojava: Democratic autonomy and women's liberation in the Syrian Kurdistan*. Pluto Press.

Kordas, O., Nielsen, J., & Neij, L. (2015). *Strategic innovation program for smart sustainable cities: Application for a strategic innovation program*. https://www.diva-portal.org/smash/get/diva2:812341/FULLTEXT01.pdf

Kovel, J. (2007). *The enemy of nature: The end of capitalism or the end of the world?* (2nd ed.). Zed Books.

Laclau, E., & Mouffe, C. (2001). *Hegemony and socialist strategy: Towards a radical democratic politics*. (2nd ed.). Verso.

Lamant, L. (2016). *Squatter le pouvoir. Les mairies rebelles d'Espagne*. Lux.

Lauriault, T. P., Bloom, R., & Landry, J.-N. (2018). *Open smart city guide 1.0*. Open North. https://www.opennorth.ca/publications/#open-smart-cities-guide

Lê-Huu, V. (2018, 11 June). Québec a maintenant sa monnaie locale. *Le Devoir*.

Leonard, J. (2018, 15 January). Barcelona to ditch Microsoft in favour of open source software. *The Inquirer*. https://www.theinquirer.net/inquirer/news/3024394/barcelona-to-ditch-microsoft-in-favour-of-open-source-software

Lubell, M. (2015). Collaborative partnerships in complex institutional systems. *Current Opinion in Environmental Sustainability, 12*, 41–47. http://doi.org/10.1016/j.cosust.2014.08.011

Magnaghi, A. (2005). *The urban village: A charter for democracy and local self-sustainable development*, Zed Books.

Magnaghi, A. (2014). *La Biorégion urbaine. Petit traité sur le territorie bien commun*. Eterotopia.

Marsal-Llacuna, M.-L., Colomer-Llinàs, J., & Meléndez-Frigola, J. (2015). Lessons in urban monitoring taken from sustainable and livable cities to better address the smart city initiatives. *Technological Forecasting and Social Change, 90*, Part B, 611–622. http://doi.org/10.1016/j.techfore.2014.01.012

Meadows, D. H., Meadows, D.L., & Randers, J. (2004). *Limits to growth: The 30-year update*. Chelsea Green.

Morozov, E. (2013). *To save everything, click here: Technology, solutionism, and the urge to fix problems that don't exist*. Allen Lane.

Morozov, E., & Bria, F. (2018). *Rethinking the smart city: Democratizing urban technology*. Rosa Luxemburg Stiftung.

Nylund, K. (2014). Conceptions of justice in the planning of the new urban landscape: Recent changes in the comprehensive planning discourse in Malmö, Sweden. *Planning Theory & Practice, 15*, 41–61. https://doi.org/10.1080/14649357.2013.866263

Offe, C. (2009). Governance: An "empty signifier"? *Constellations, 16*(4), 550–562. https://doi.org/10.1111/j.1467-8675.2009.00570.x

Ouellet, M. (2006). Le smart growth et le nouvel urbanisme: Synthèse de la littérature récente et regard sur la situation canadienne. *Cahiers de géographie du Québec, 50*(140), 175–193. http://doi.org/10.7202/014083ar

Owen, D. (2012). *The conundrum: How scientific innovation, increased efficiency, and good intentions can make our energy and climate problems worse.* Riverhead.

Pateman, C. (2012). Participatory democracy revisited. *Perspectives on Politics, 10*(1), 7–19. https://doi.org/10.1017/S1537592711004877

Pearce, D. W., & Atkinson, G. D. (1993). Capital theory and the measurement of sustainable development: An indicator of weak sustainability. *Ecological Economics, 8,* 103–108. https://doi.org/10.1016/0921-8009(93)90039-9

Peck, J. (2015). *Austerity urbanism: The neoliberal crisis of American cities.* Rosa Luxemburg Stiftung.

Peck, J., Theodore, N., & Brenner, N. (2009). Neoliberal urbanism: Models, moments, mutations. *SAIS Review of International Affairs, 29*(1), 49–66. http://doi.org/10.1353/sais.0.002.

Pereira, G. V., Cunha, M. A., Lampoltshammer, T. J., Parycek, P., & Testa, M. G. (2017). Increasing collaboration and participation in smart city governance: A cross-case analysis of smart city initiatives. *Information Technology for Development, 23*(3), 526–553. https://doi.org/10.1080/02681102.2017.1353946

Perng S.-Y., Kitchin, R., & MacDonncha, D. (2017). Hackathons, Entrepreneurship and the passionate making of smart cities. The Programmable City Working Paper 28.

Reyes, O. (2017, 16 August). Fearless cities: The new urban movements. Red Pepper. https://www.redpepper.org.uk/fearless-cities-the-new-urban-movements/

Sauter, M. (2018). Google's guinea-pig city. *The Atlantic.* https://www.theatlantic.com/technology/archive/2018/02/googles-guinea-pig-city/552932/

Schlosberg, D. (2007). *Defining environmental justice: Theories, movements, and nature.* Oxford University Press.

Scholz, T., & Schneider, N. (2016). *Ours to hack and to own: The rise of platform cooperativism: A new vision for the future of work and a fairer internet.* OR Books.

Servigne, P., & Stevens, R. (2015). *Comment tout peut s'effondrer. Petit traité de collapsologie à l'usage des générations présentes.* Seuil.

Sidewalk Labs. (2018). "Sidewalk Toronto." https://sidewalktoronto.ca/

Smart Cities India. (2015). *Smarter solutions for a better tomorrow.* https://aquafoundation.in/pdf/Smart_Cities_India.pdf

Termeer, C. J., & Bruinsma, A. (2016). ICT-enabled boundary spanning arrangements in collaborative sustainability governance. *Current Opinion in Environmental Sustainability, 18,* 91–98. http://doi.org/10.1016/j.cosust.2015.11.008

United Nations Environment Programme (UNEP). (2002). *Melbourne principles for sustainable cities.* UNEP.

Ville de Grenoble. (2017). *Guide pratique de la participation citoyenne à Grenoble*, Direction de projet démocratie locale. https://www.oidp.net/docs/repo/doc975.pdf

Ville de Montréal. (2018). *Accélérer Montréal: Stratégie de développement économique 2018-2022*. http://ville.montreal.qc.ca/pls/portal/docs/PAGE/AFFAIRES_FR/MEDIA/DOCUMENTS/ACCELERER_MONTREAL.PDF

Wallace-Wells, D. (2019). *The uninhabitable earth: Life after warming*. Tim Duggan Books.

Wilcox, D. (1994). *The guide to effective participation*. Partnership Online. http://ourmuseum.org.uk/wp-content/uploads/The-Guide-to-Effective-Participation.pdf

Williams, A., Goodwin, M., & Cloke, P. (2014). Neoliberalism, big society, and progressive localism. *Environment and Planning, 46*(12), 2798–2815. http://doi.org/10.1068/a130119p

World Commission on Environment and Development (WCED). (1987). *Our common future: Report of the World Commission on Environment and Development*. Oxford University Press.

Wright, E. O. (2010). *Envisioning real utopias*. Verso.

2 Mobilisons-Nous: "Violent Infrastructure" and Pedestrian Space in Montreal

KEVIN MANAUGH AND NATALYA BEREZINA DRESZER

City life is subtly but profoundly changed, sacrificed to that abstract space where cars circulate like so many atomic particles.

Henri Lefebvre (1991, p. 312)

Introduction

Rodgers and O'Neill describe "infrastructural violence" as a means to "nuance our analyses of the relations between people and things that can converge daily in urban life to the detriment of marginalized actors" (Rodgers & O'Neill, 2012, p. 403). Perhaps no feature of modern cities better exemplifies the concept of "infrastructural violence" than urban highways with their on- and off-ramps, overpasses, and flyovers generating noise, vibrations, and pollutants affecting the health, well-being, and accessibility of local residents. In many cases, adding insult to injury, the populations exposed to the harms of these projects do not reap the benefits (in the form of faster mobility or convenient accessibility to amenities). While Robert Moses's legacy of urban highways through low-income and minority neighbourhoods are the most well-known example, these decisions are not unique to New York City. Many parts of cities throughout the world have suffered a similar fate when the "desire lines" of transport engineers ran through poor and marginalized neighbourhoods.

The negative environmental, health, and social impacts of automobile dependency are well known. Canada has among the highest rates of road length per capita in the world, with well-established links to increased energy use, habitat fragmentation, and air pollution (including GHG emissions as well as particulate matter and other pollutants) (Statistics Canada, 2006). Increasing active transportation (walking and cycling) also has well-established health and environmental benefits. Walkable space is increasingly being seen as a powerful means to address multiple societal goals, including reducing greenhouse gas emissions and congestion, improving public health, and fostering social inclusion. In Canada, where the share of walking trips declined from 7.0 per cent in 1996 to 5.7 per cent in 2011 and obesity rates continue to rise (Twells et al., 2014), a variety of stakeholders from public health and provincial and federal transportation agencies have begun advocating for an increase in walking as an everyday mode for utilitarian trips (Public Health Agency of Canada, 2014; Transport Canada, 2011). While pedestrian and cyclist injuries and fatalities dropped between 2011 and 2015, they remain high (Transport Canada, 2015).

Though increasingly lauded as a solution to many of the environmental, social, public health, and aesthetic problems associated with the car-dependant urban form, the concept of the "walkable city" is, obviously, not a new one (perhaps turning the concept of innovation on its head). For millennia the size and scale of urban development was constrained by human walking speed. This technical limitation meant that most pre-industrial cities fit a dense mix of residential, commercial, religious, and institutional buildings into an area within a half mile of a central square (Southworth, 2005). Each subsequent advance in transport technology led to a gradual horizontal stretching of urban space as streetcars and trains greatly expanded the speed and reach of travellers, allowing ever-increasing distances between homes and workplace and leisure destinations (Sheller and Urry, 2000). Until the mid-twentieth century, North American cities developed relatively densely along transit corridors connecting residential and commercial districts (Southworth, 1997). As the car gained prominence and political support, the shape of cities began to dramatically change. Mutually reinforcing legal, social, and built environment factors began to favour and prioritize urban space for vehicles rather than people (Shill, 2019). These factors include single-use zoning, initially conceived to reduce exposure to industrial pollution by separating residential and industrial land uses (Village of Euclid v Ambler Realty Co., 1926). Pedestrian behaviour was also increasingly regulated, both through the formation of social norms and the codification of laws and regulations. For example, the criminalization of "jaywalking" was introduced in the early 1920s, to curtail the rise in pedestrian injuries and death as a result of vehicular collision. However, as a sign of things to come, this victim-blaming approach prioritized the speed and convenience of drivers at the expense of pedestrians (Norton, 2007). A major loss in this battle over street space was arguably the passage of the $26 billion highway program in the United States (Mumford,

1964), which ushered in the age of highways within cities, as opposed to a means to connect cities to one another, i.e., the "townless highway."

Moreover, the positive and negative impacts of pedestrian networks and space are not equally distributed across society. Low-income neighbourhoods in Canada's largest cities bear a disproportionate number of pedestrian-motor vehicle collisions, including those involving children, partly due to the presence of high-speed arterial roadways, on- and off-ramps, and other built environment features that do not favour walking (Morency et al., 2012; Yiannakoulias & Scott, 2013). Low-income carless individuals may be left isolated from amenities related to health, education, food, and affordable housing, which are necessary for inclusion and well-being, and thus compelled to walk longer distances than more advantaged counterparts would tolerate (Manaugh & El-Geneidy, 2012). Finally, while many cities throughout the world have been investing in high-profile pedestrian "mega-projects" such as New York's High Line and Philadelphia's Reading Viaduct, these projects have also been criticized for leading to unaffordability of housing units as well as questions about race, space, access, and inclusion.

Poor environments for active transportation may perpetuate disadvantage along social and geographic lines. Even more so than automobile and transit modes of transport, walking has clear and immediate links to social inclusion and wider ideas of the "right to the city." With its roots in the right to access and shape urban spaces (Lefebvre, 1991), social inclusion enables groups to participate and prosper in society, while being equitably considered in the distribution of impacts from transportation decision-making (Jones & Lucas, 2012). Some transportation justice scholars argue that planning should draw from ethical principles to improve transportation opportunities for socially excluded groups as a means to augment their capabilities (Martens, 2016; Pereira et al., 2017).

Public participation in transport planning (McAndrews & Marcus, 2015) has been seen as a way to bring into practice the intersection of planning and social inclusion. Arnstein (1969) noted that participatory exercises can range from manipulation and tokenism to citizen control, a theme also noted by more contemporary scholars (Monno & Khakee, 2012). Importantly, socially disadvantaged groups may not be well-represented in public meetings, petitions, and other forms of participation as a result of their lack of access to information compared to other groups, whether it be cultural and language barriers, lack of free time, or availability of transport (Hodgson & Turner, 2003). Synthesizing themes of pedestrian space, participation, and the right to the city, Soja writes, in reference to Lefebvre's "right to the city" idea, of "the need for those most negatively affected by the urban condition to take greater control over the social production of urbanized space" (Soja 2010, p. 6).

Montreal has recently undertaken several large-scale transportation infrastructure projects. These include the reconstruction of the Turcot Interchange (the congruence of three urban highways near the city's downtown) and replacing the Jacques Cartier Bridge (the longest of the four bridges in Montreal that span the St. Lawrence River). Meanwhile, the Reseau Express Montreal (REM) electric light rail train line is under construction, after a surprisingly quick approval. Extending the "Blue" line of the Metro is getting strong support from Ottawa, Quebec City, and City Hall, and the "Pink" Metro line, championed by Montreal Mayor Valérie Plante in her successful election campaign in 2018, is also being considered. At the same time, the City of Montreal is increasingly looking to pedestrian-oriented projects as a component of a move towards sustainable mobility.

This chapter examines three projects in Montreal: the "Dalle-Parc," which would provide pedestrian and cyclist access across the rebuilt Turcot Interchange between the neighbourhoods of Notre-Dame-de-Grâce and the Southwest; Green Alleys; and the Sentier Petit Bourgogne. It also touches on the Ste-Catherine pedestrian-oriented street redesign and the $50 million "Mountain to River" promenade. In the context of pedestrian planning in Montreal, this chapter explores agency and power, and the capacity of local residents to influence infrastructure decisions that affect their well-being, through participatory and advocacy mechanisms, with varying degrees of effectiveness. We examine how Montreal struggles to provide safe walkable environments in its move towards sustainable infrastructure, and in particular infrastructure that provides for walkability, which contributes to a balance of environmental, economic, and social interests. Two of the case studies deal explicitly with bringing marginalized voices to the table, while the third case study, Montreal's Green Alleys, highlights how liveability and safety issues can be brought to bear in the pedestrian realm of a wealthier (on average) neighbourhood.

The Turcot in Context

Throughout the 1960s and 1970s, Montreal undertook several large-scale infrastructure projects in transportation, housing, and spectator sports, many of which still shape the character and image of the city today (such as Habitat 67 and the Olympic Stadium). Most were directly or indirectly linked to the 1967 World Expo, for which the city strove to present a modern face to the world (as it also did for the 1976 Summer Olympics). The Montreal Metro system dates from this time, as does the Turcot Interchange, which opened in 1966 and 1967 respectively. The Turcot Interchange is located between two Montreal boroughs (the Southwest, and Notre-Dame-de-Grâce; see Figure 2.1). The neighbourhoods of St-Henri and Little Burgundy, located within the Southwest Borough, are traditionally working-class neighbourhoods near the Lachine Canal, which, before the opening of the St-Lawrence Seaway in 1959, was the sole way to ship goods from the Atlantic to the Great Lakes. The impact of the seaway and its role in reducing the importance of the canal for industrial and shipping uses, along with the suburbanization of industry, is connected to the loss of roughly 40 per cent of the area's manufacturing jobs (Centre for Oral History and Digital Storytelling, 2013). After the complete closure of the Lachine Canal to ship traffic in 1970, the once economically vibrant communities further suffered, with the population of the area dropping from 120,000 to 66,000 between 1961 and 1991. Also, in a somewhat ironic touch, the Turcot Interchange was built at its elevated height to accommodate the passage of ships on the canal, making the height unnecessary just a few short years later.

The Turcot Interchange connects three highways and is part of larger east-west and north-south highway corridors. The original construction of the east-west corridor resulted in the displacement of 15,000 people and the bulldozing of 3,300 homes and was funded as a result of municipal, provincial, and federal governments agreeing to integrate the project into the Trans-Canada Highway system (Poitras, 2011). The selection of the location involved negotiating with elements of physical geography (including the St. Jacques Escarpment and location of the Lachine Canal), as well as political, school district, and religious jurisdictional boundaries. For example, the Montreal Catholic archdiocese had a stake in whether their parish boundaries would be affected. In addition, railway rights of way as well as heritage buildings further to the south of the site had to be taken into account. The Turcot Interchange was originally designed to carry 50,000 to 60,000 vehicles per day, but carried up to 300,000 at its peak. The disruption it caused for the surrounding neighbourhoods still affects the communities today.

By the early 2000s, much of the infrastructure built in the 1960s in Montreal was deteriorating and in need of repair. Built in 1959, the downtown Park/Pins Overpass was finally demolished in 2005, after decades of activism (La Communauté Milton Parc, 2002). Constructed in 1969, the de la Concorde Overpass collapsed in 2006 (Leavitt, 2016), resulting in five deaths and six injuries. The tragedy of de la Concorde ushered in a wave of debate about crumbling infrastructure, and Quebec announced plans to reconstruct the Turcot Interchange. This massive project entails simultaneously tearing down the "aerial spaghetti" (Aubin 2009) of interlocking overpasses, on-ramps, and connections that were deemed in

"Violent Infrastructure" and Pedestrian Space in Montreal   49

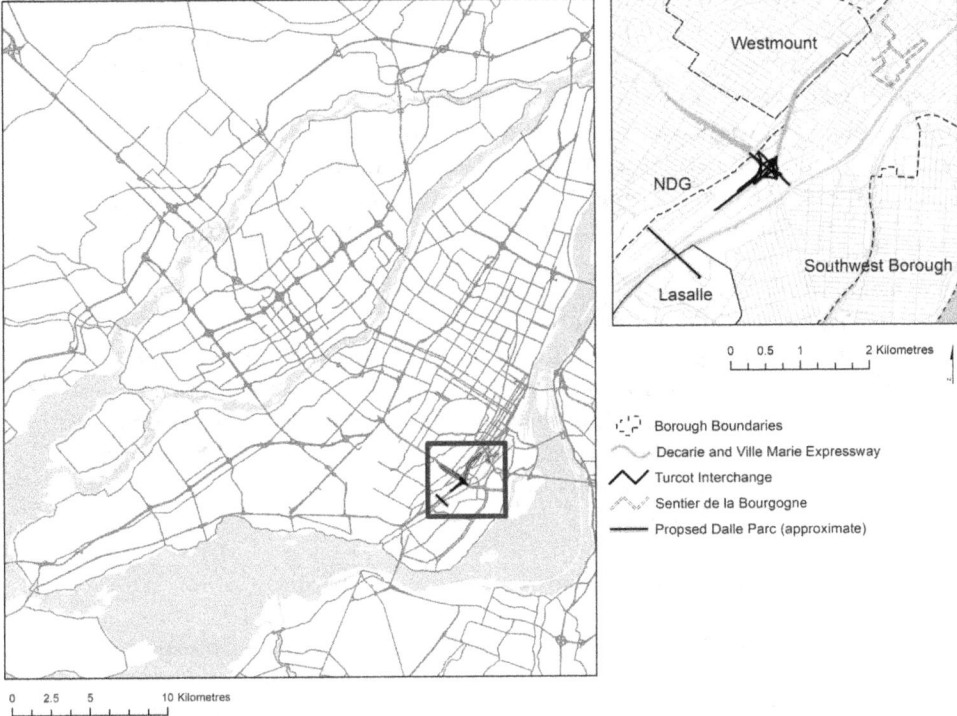

Figure 2.1 Montreal context. Map courtesy of the authors.

danger of collapse, and rebuilding a new interchange at ground level. In addition to aesthetic benefits, rebuilding at ground level would be much cheaper than rebuilding in the air. By the beginning of its reconstruction in 2010 an estimated 300,000 vehicles used the infrastructure daily, 30,000 of which were trucks, making it the busiest interchange in Quebec (Radio Canada, 2013; Transport Quebec 2009).

The Turcot Reconstruction

After the plans were announced, stakeholders took part in discussions and debates for nearly a decade. A major source of contention is the fact that, in addition to the federal government (which holds sway over the Lachine Canal, which flows beneath the structure, as well as the nearby Champlain Bridge across the St. Lawrence River), five levels of government are involved in planning transportation infrastructure in Montreal: the Province of Quebec, the Communauté métropolitaine de Québec, the agglomeration of Montreal, the City of Montreal, and the individual boroughs of the city (Fischler, 2009). Of particular importance is the provincial government, which has ultimate control over highway oversight, and the City of Montreal, which can influence the decision-making of the provincial government, and the borough governments, which can work on smaller scales with their citizens (Gauthier et al., 2009).

Another complication is the fact that the project has seen tremendous institutional changeover, with five different Montreal Mayors, eight Ministers of Transport, three Quebec Premiers, and a variety of other shifting stakeholders. Through a process of approvals and disapprovals, rewrites, and consultations, disparate voices have been heard and incorporated into the changing plans for the structure. Local governments and community groups have formed multiple coalitions over the decade, and political powers fought while striving to find consensus on a way forward. The projected cost of the Turcot has doubled since its original proposal, currently at $4 billion, and completion of the project took years longer than originally planned.

Stakeholders and Background

Before rebuilding the overpass, the Quebec Ministry of Transport (the MTQ, in its French acronym) went through a legally mandated review. Between 2009 and 2010 the MTQ attempted to obtain approval from the Environmental Assessment Board (French acronym, BAPE) (Développement durable, 2018). BAPE rejected plans proposed by the MTQ multiple times because they included demolishing 160 housing units and expanding automobile infrastructure without providing adequate public transportation or transportation alternatives (CBC News, 2010a; Transport Quebec, 2009). The MTQ's primary focus was to relieve traffic congestion through these plans and address the 300,000 vehicles that travelled through the Turcot daily. BAPE heard from over 90 citizen groups about the Turcot Interchange (CBC News, 2010a), and during that time, these groups came together to provide input and insight that they hoped would be taken into account before the Turcot was constructed. The message of these groups was the importance of undoing the damage that had been done with the infrastructure in the 1960s. The overarching theme of multiple groups was to make the interchange smaller and less costly and less socially and environmentally impactful. In late November 2009 the MTQ proposal was officially denied.

One of the most important ways that citizens made their voices heard during this time was the publishing of a book of essays by over 20 contributors, titled *Montréal at the Crossroads: Superhighways, the Turcot and the*

Environment (Gauthier et al., 2009). The articles include background information about the Turcot's history, proposed solutions for the Turcot that promote a more socially and environmentally sustainable infrastructure project, and what the Turcot could mean for people in Montreal. Contributions came from professors and students from Montreal universities as well as community members.

In 2010, BAPE approved the plans of the MTQ after the MTQ incorporated most of the recommendations suggested by BAPE, including decreasing the number of housing units destroyed and creating a pedestrian overpass connecting two boroughs called the Dalle-Parc, which would provide an accessible link between two areas of Montreal that would otherwise take three times as long to access and would include several dangerous intersections.

BAPE's decision to approve the plans of the MTQ was nothing if not controversial. Leaders of Project Montreal, at the time an up-and-coming political party from the progressive Plateau borough, referred to the plans as "scandalous." Citizens and community groups were outraged (CBC News, 2010a; CTV News Montreal, 2010). The Mayor, Gerald Tremblay, who later resigned for suspected fraud (Gyulai, 2018), released an alternative plan to the MTQ for the Turcot that decreased the number of cars, eliminated destruction of buildings and infrastructure in surrounding boroughs, and established lanes dedicated to public transportation. The plan was praised by community leaders and politicians alike, but was not enough to change the provincial plans (CBC, News, 2010a).

One of the largest umbrella organizations that mobilized activism surrounding the Turcot at this time was Mobilisation Turcot. They organized public events such as a march through the Plateau neighbourhood while dressed in 1950s clothing to symbolize the antiquated plans of the Turcot Interchange, since it was being planned to increase automobile accessibility without thought for public transportation or active transportation (Mobilisation Turcot, 2011). Mobilisation Turcot also drew up a counterproposal based on the work of Montreal urban planning professors and scholars and presented it to Quebec's Minister of Transport. The alternative proposal was projected to cut the expense of the Turcot in half and take less than two years to complete, while addressing the traffic congestion that the MTQ was concerned about (Mobilisation Turcot and le Groupe de recherche urbain Hochelaga-Maisonneuve [GRUHM], 2012). In addition to the protest and counterproposal, Mobilisation Turcot submitted a letter to the Ministry of Transport signed by hundreds of citizens and groups in and around Montreal "asking for a simpler and more ecological project for the Turcot Interchange" (Mobilisation Turcot, 2013a). They argued that the Minister of Transport was wrong in announcing that the MTQ's plans respected provincial laws, which stipulate that infrastructure projects must work to reduce environmental impacts, and the Turcot would, in fact, increase environmental impacts.

Despite all the organizing to change the plans for the Turcot, in 2013 the new Mayor of Montreal, Michael Applebaum (who later resigned in response to 14 criminal charges and his subsequent his arrest; Bruemmer, 2017) stood in favour of the plans for the Turcot, and most members of the movement felt nothing more could be done to resist. All of the mobilizing and organizing had little or no effect without political backing, and it seemed that the MTQ was accountable to its citizens only through BAPE. One last straggling protest was made that year, a yarn-bombing knitted graffiti of the Turcot in solidarity with several community groups (Mobilisation Turcot, 2013b; TRICOT, 2013). While this last protest was given prominent coverage, no amount of media attention mattered without support from city government.

Construction and Removal of Pedestrian Access

In late 2016, after demolition and reconstruction had begun, the MTQ removed plans for the Dalle-Parc from its website. Felix Gravel of the Regional Environmental Council of Montreal (CRE-Mtl) wrote an open letter co-signed by almost 60 organizations and individuals to the Quebec Premier, expressing outrage at the removal of the Dalle-Parc from the plans. The letter explained that the MTQ was given permission to start building because the project was promised to include the Dalle-Parc (Gravel, 2016). After months of silence, the MTQ confirmed that there were no longer plans to build the Dalle-Parc in the Turcot Interchange. This announcement frustrated and angered politicians and individuals in Montreal and breathed new life into mobilizing protests against the Turcot Interchange (Gould, 2017).

CRE-Mtl took a role similar to Mobilisation in the earlier days of the Turcot as an umbrella organization for all individuals and groups that wanted to mobilize to protest the disappearance of the Dalle-Parc. Under CRE-Mtl, borough councils, medical organizations, pedestrian and bicycle groups, and individual citizens joined in protests and petitions to turn the political tide to create real action for the Dalle-Parc. CRE-Mtl called on citizens to draw how they imagined the connection might look between two neighbourhoods across the Turcot (Mobilisons-Nous! 2017b), organized a rally and demonstration along the Lachine Canal at the proposed location of the Dalle-Parc (D'Amours, 2017), organized a cycling demonstration with the detour route they would have to take to reach the points that would be connected by the Dalle-Parc (Cambron-Goulet & Shaffer, 2018; Conseil régional de l'environnement de Montréal, 2017a, 2017b, 2017c; D'Amours, 2017; Mobilisons-Nous! 2017b; Montreal, 2017; Nadeau, 2017), and circulated a petition to the Quebec Premier to maintain the Dalle-Parc, receiving 11,000 signatures (De Rosa, 2017; Mobilisons-Nous! 2017b).

After an entire year of this mobilization, more political news was released about the Turcot: an MNA questioned the new Minister of Transport for Quebec on the Dalle-Parc and the minister promised that the MTQ would leave space for the Dalle-Parc and that they would discuss the Dalle-Parc further (Symon, 2017). At about this time, Valérie Plante, a long-time member of Project Montreal, was elected Mayor. She threw in the support of the Mayor's Office by announcing a $125,000 feasibility study to be paid for by the City of Montreal (CBC News, 2018).

The backing of this movement by the Montreal municipal government implies reason for greater hope this time, and the government backing is no coincidence. While the governing bodies are separate from their citizens, the citizens can ultimately decide on the representation they would like. In the Plateau, Project Montreal has held all seats on the council for nearly a decade, and the same party has now spread to the Mayor's Office. As of this writing, the Dalle-Parc is set to move forward.

The Importance of Borough Governments

Borough councils have been at the heart of resisting the Turcot reconstruction since the plans were first introduced. And although local governments have minimal power over provincial highway plans, they can have local effect. In 2017, CRE-Mtl hosted a consultation open to all people and organizations. The consultation focused on the urban neighbourhoods surrounding the Dalle-Parc. That consultation aided consultations between CRE-Mtl and municipal politicians. An "Urban, Economic, and Social Development Plan" was adopted on 28 May 2018 by the City of Montreal and, while far from perfect, does seek to ameliorate the quality of life for neighbourhood residents and address some concerns raised in the consultations (Gravel, 2018; Le Sud-Ouest Montréal, 2018; Le Sud-Ouest Montréal & Transport, 2018).

Sentier Petit Bourgogne

Just east of St-Henri and in the shadow of the Turcot Interchange is the neighbourhood of Little Burgundy within the Southwest Borough (Figure 2.1). Lying along the east-west axis that felt the brunt of 1960s modernism, this historically Black neighbourhood was once a bastion of jazz before the bulldozers tore through the area in the 1960s in a wave of urban renewal, which included the Ville-Marie Expressway and the demolition of several residential blocks and the altering of the street grid to favour wide, high-speed arterial roads (Poitras, 2011). This area, perhaps more than any other Montreal neighbourhood, exemplifies "Haussmannization" through infrastructural violence, which has been observed throughout the world (Rodgers, 2012).

In this lower-income community, many residents depend on active and public transport for daily mobility. However, the area is not conducive to such movement. In 2013, the Little Burgundy Coalition formed to use active transportation to rediscover the historical and cultural richness of the neighbourhood. Through a walking trail, redesign and improvements in public space, and a cycling path, the coalition aims to honour and celebrate the neighbourhood's past with a modern vision of active mobility. This project is within the purview of Quartier 21. This initiative, begun in 2005, is based on the UN Agenda 21, which encourages local sustainable development projects, and seeks to empower local residents by including them in the planning and implementation of local projects. The path will link two Metro stations and connect locally important sites such as a community garden and Parc des Jazzmen, named in tribute of the history of jazz in the area. Furthermore, local artists have created murals along the walking path that celebrate the local history through images of local jazz luminaries such as Oscar Peterson and Oliver Jones. They also pay homage to the former rail line and its importance in the employment of former residents who were denied work in other industries and areas of the city.

Construction of the walking path includes safety measures at intersections and along paths, greening, and protection and enhancement of the natural environment. Over 600 residents have made their voices heard in consultations focused on seniors and youth. Beginning as a grass-roots, community-led project, the *sentier* has secured multiple partners and funding sources. For example, both the City of Montreal and the Montreal public health agency have offered funding, through the Quartier 21 initiative as well as the South West borough.[1] By partnering with established entities, local voices and plans are being implemented and brought to fruition.

---

1 Other partners and funding sources include CSSS Sud-Ouest-Verdun, Éco-Quartier Sud-Ouest, Office municipal d'habitation de Montréal, Regroupement économique et social du sud-ouest, Table de concertation des organismes en enfance-famille de St- Henri et Petite-Bourgogne, Regroupement des organismes pour aînés et aînées du sud-ouest de Montréal, and Équipe mobile alimentation.

The Alleys of Montreal

Much of the residential built form of Montreal is characterized by alleys that are open to car traffic and are sometimes used to access backyard parking and provide shortcuts. These most often consist of a long "north-south" axis, with two shorter perpendicular connections allowing four entrances to the alley. The first alleys appeared in wealthier Montreal neighbourhoods in the 1840s to provide access to the rear of buildings (Robert, 2014). Stables, laundry lines, and coal sheds were located in the alleys, which hid them from public sight (Weaver Crawford, 2015). But with the shift from carriages to cars and coal heating to hydro, alleys saw a decrease in their intended use.

With the lack of space for children to play within the city, alleys became informal playgrounds, especially after the Second World War. Neighbours met here, and they became a place for street vendors and fuel oil delivery. But alleys were paved in the 1960s and the park-like aspects of these public spaces were removed (Weaver Crawford, 2015). Activities diminished in these areas for lack of lighting and upkeep, and alleys gained a reputation of being dangerous, so they were avoided by city residents.

Mayor Jean Drapeau implemented two programs in the late 1980s to improve these spaces: Place au Soleil and Operation Tournesol. Place au Soleil provided grants for neighbourhoods willing to transform their alleys into small parks (Wise, 2004). While the program ended in 1988, it converted 58 alleys across Montreal into parks. These parks are often closely guarded by surrounding residents and are seen as private property, not as public space (Wise, 2004). Operation Tournesol was initiated in 1989 to increase sunlight in alleys by demolishing coal sheds no longer in use. It resulted in the destruction of 35,000 coal sheds in Montreal (Wise, 2004). The overall goal of Drapeau's programs was to improve water retention and security conditions (Weaver Crawford, 2015). While they ameliorated the infrastructure of alleys, these programs were top-down in their implementation and did not focus on building public community spaces.

Green Alleys

Green Alleys (*ruelles vertes*) have been converted into green spaces by removing asphalt and planting vegetation. Green alleys are associated with local environmental impacts such as reduction in urban "heat islands," storm water retention, and increased biodiversity, as well as an increase in spaces for social interaction and play, and, at least symbolically, a move away from automobile-centric spaces. In Montreal, these *ruelles vertes* can be informal or officially recognized and supported by local boroughs. In officially sanctioned alleys, borough councils provide resources for citizens to create and implement green alleys once they have a plan and have created an alley greening committee. In contrast, unofficial green alleys serve the same purpose but do not have official support from the city.

The concept of unofficial green alleys started in the late 1980s when community members reclaimed unused space around their homes (Le Plateau Mont-Royal, 2016). Multiple stakeholders and their motivations guided the development of green alleys in Montreal, including local architecture students enthusiastic about transforming space (Plourde-Archer, 2013), and communities' desire to reclaim underutilized urban space. The first green alley in Montreal was recognized in 1997 in Plateau (Le Plateau Mont-Royal, 2016). Although government support provided financial and political benefits to the green alley movement, non-official green alleys remain sites of political tension. Formalization of green alleys has contributed to debates about citizen voices in decisions, resulting in a political, physical, and social transformation of space and the role of citizens in planning.

Alleys in Montreal have limited maintenance and by-laws dictating their use. The City of Montreal owns only 38 per cent of alleyways within its jurisdiction: 46 per cent are privately owned (mainly by Hydro Québec), and 16 per cent are classified as private-declared public (Wise, 2004). In the Plateau there are strict regulations on the appearance of building front; however, alleys lack the same requirements. Sufficient clearance for emergency vehicles is now the only public provision provided by alleys; they are not wide enough to allow for garbage trucks (Wise, 2004). Additionally, the City of Montreal does not provide cleaning and snow removal. The latter occurs only if at least one-third of residents on a block ask for it (Wise, 2004). The informality of alleys as compared to streets has resulted in this transformation from service route, to under-used and dangerous, to an opportunity for creating community spaces.

Two Examples: "Burner Alley" and "Esplaneuve"

Since green alleys were legally recognized by borough councils in the 1990s, borough governments supported residents in creating official green alleys and finding expertise in planning and creating them (Ville de Montréal, 2015). Community members whose residences were connected directly to the alley had to join and petition for green alley conversion. A majority of residents had to sign on to the application, and then the borough government would assess the application (Ville de Montréal, 2015). Once the borough established that the alley was a good candidate for a green alley, the borough would remove the pavement and provide materials for residents to create their green alley: soil, compost, plants, woodchips, etc. (Ville de Montréal, 2015). These resources can facilitate more extensive transformations than those undertaken by residents alone.

In 2009 Projet Montreal won all seven seats on Le Plateau borough council. They ran on a platform of increasing green spaces and decreasing automobile traffic in the neighbourhood, focusing on small projects such as *"les sallies de trottier"* and *"ruelles vertes,"* or sidewalk extensions and green alleys. They reversed the burden of proof from residents who want the alley to residents who do not: now residents must only form a green committee that will oversee the alley and file an application. If residents wish to prevent creation of an alley, they must file a petition arguing that the majority of residents do not want it. Once the alley is approved, the borough provides resources for residents to create the alley, such as soil and native plants. One such alley is "Esplaneuve," created by residents at the corner of Jeanne-Mance and Esplanade.

Residents' hands-on initiative to create this fostered new relationships and even changed the minds of some anti-alley residents to being enthusiastic supporters. The creation of green committees has sometimes spurred political involvement in the community. The Esplaneuve Green Committee also became more politically involved in 2018, officially supporting the city's plan to close Mount-Royal to through traffic (Scott, 2018). They also signed on to Matthew Chapman's petition to the city to remove 3 per cent of paved area in Montreal per year, resulting in a 50 per cent decrease in paved area by 2050 (Chapman, 2018). Support from the borough and municipal government for projects like the Esplaneuve Green Alley helps make the city more pedestrian-friendly and also unleash citizen energy about city initiatives.

Not all stories of green alleys are simple and straightforward. One alley, affectionately named "Burner Alley" for the large number of its Burning Man participants, was almost completely bulldozed because it did not have the correct city permit. The space was created as community space made by residents for residents without official help or recognition from the borough, but over the course of a decade, residents surrounding Burner Alley worked to beautify it through art installations, plants, and furniture (Birkbeck, 2015). In September 2015, city bulldozers showed up unannounced. Local residents blocked the bulldozers and tweeted what was happening. That caught the attention of a City Councillor, who made his way down to the scene and helped mitigate the tension between the city government and its residents. Soon after, Burner Alley was upgraded to a recognized "green alley."

The story of this alley highlights the tensions that surround the way groups view their rights to public space. Residents took over an unused alley to create a sanctuary away from cars. Because their action was not backed by the government, it was seen as a hostile attempt to occupy public space, and the city threatened to bulldoze it. Similar to the larger-scale example of the Turcot, residents had little power over the city's choices until they had political backing. Once the City Councillor recognized the plight of Burner Alley, residents gained official recognition. The attempted removal of Burner Alley offers a reminder that despite a stated desire to promote green alleys, they must be approved by the borough.

In 2016, after a 6-year-old child was hit by a truck coming out of an alley in the Plateau, the borough promised to double the number of alleys closed to traffic from 35 to 70 by 2018. Additional safety measures have also been put in place, such as *"saillies de trottoir,"* which remove eight parking spaces from an intersection and extend the sidewalks. These changes force cars to slow down, make the intersection more visible, and decrease the distance to travel between one side of the street and the other (Projet Montreal 2018). In 2018 the Plateau borough government aided in the creation of 12 more Green Alleys, and was providing free compost, plants, woodchips, and vines to all residents of the Plateau. Government backing of pedestrian spaces is mobilizing and implementing these spaces at a speed that far exceeds the efforts of residents without political backing.

Discussion

While varied in scale and scope, these examples shed light on how Montreal residents are pushing back against the automobile-centred paradigm that has dominated city planning and transportation decision-making for decades. They all touch on important and under-studied elements in the overlap of environmental and social inclusion goals. Provision of walkable urban space is much more than a public health issue or a question of low-carbon transport. It is integral to understanding wider issues of social and spatial justice and addressing the question about what kind of cities we, as a society, want.

These examples look at spaces that have been traditionally allocated to automobiles at the expense of pedestrians and cyclists and explore how residents' efforts can successfully lobby for change in street space.

There are important differences: green alleys are not responding explicitly to oppressive infrastructure to the same extent that the other projects are, but these lessons are of interest across projects. While we cannot demonstrate what might have happened under different circumstances, we can draw some important lessons from these examples. We see the importance of grass-roots organizing and activism, and of garnering official, political support for a cause. This brings up important distinctions raised by David Harvey in *Rebel Cities* (2013) as well as the key role of collaboration among stakeholders. It is difficult to imagine the Dalle-Parc, Sentier Petit Bourgogne, or officially recognized green alleys without the combined efforts of dedicated individuals and groups getting politicians and decision-makers on board.

One of the most useful insights that can be gained from the story of the Turcot Interchange is that checks and balances can work only if the government is accountable. The environmental review (BAPE) mandated by the provincial government seemed to force the MTQ to listen to Montreal citizens and compromise on environmental and social issues by decreasing the number of housing units to be demolished and providing access for pedestrians and cyclists to cross the interchange. However, because the MTQ could later dismiss these compromises and remove the Dalle-Parc from the plans without consulting BAPE, there was no accountability to ensure BAPE's influence in the project. Only community-led petitions and activism brought the project back to the drawing board.

Despite failed consultations with the provincial government, borough consultations tended to succeed. Smaller-scale community-based discussions such as the one with CRE-Mtl and the boroughs around the Turcot tended to result in more tangible policy outcomes, perhaps because borough governments listened to the information collected through these deliberations and tried to find solutions for citizens within the bounds of their power over the Turcot.

It is worth mentioning that other community groups such as Transport Actif Parc-Extension are also trying to address and remediate unwalkable environments. These efforts to improve liveable urban space work on a much smaller scale than the Dalle-Parc but are important in advocating a safe, inviting, and practical pedestrian environment. Elsewhere in Montreal, several other projects are under consideration. The Mountain-to-River Project plans to make a pedestrian path between the Mountain and the Old Port of Montreal. Construction is already well under way, with an area from McTavish to the Mountain complete. Another project under consideration is the St. Catherine's pedestrianization project (City of Montreal, 2018a, 2018b; CTV News, 2018; Shaffer, 2017). This project was born from the collaboration of L'Alliance pour un nouveau partage de la rue Sainte-Catherine and the city government. Started in 2014, L'Alliance comprises individual citizens and organizations such as CRE-Mtl that share a vision of pedestrian accessibility in Montreal. CRE-Mtl is also working with the city to implement more Vision Zero and Universal Accessibility infrastructure around the city. All of these projects and initiatives have the potential to address environmental sustainability while taking into account the voices of local residents and keeping social inclusion in mind.

REFERENCES

Arnstein, S. R. (1969). A ladder of citizen participation. *Journal of the American Institute of Planners, 35*(4), 216–224. https://doi.org/10.1080/01944366908977225

Aubin, H. (2009). Wrong road forward: The Turcot project fails to do much to accommodate trains and other public transit. In P. Gauthier, J. Jaeger, & J. Prince (Eds.), *Montreal at the crossroads: Superhighways, the Turcot and the environment.* (pp. 7–8). Black Rose Books.

Birkbeck, T. (2017, 14 April). Burner Alley in Plateau-Mont-Royal dismantled to residents' dismay. CBC News.

Bruemmer, R. (2018, 17 April). Michael Applebaum resurfaces as N.D.G. real estate agent. *Montreal Gazette.*

Cambron-Goulet, D., & Shaffer, M.-È. (2018, date). La dalle-parc Turcot renaît de ses cendres. *Métro.*

CBC News. (2010a, 12 March). Turcot reconstruction plans changed.

CBC News. (2010b, 21 April). City has new vision for Turcot interchange.

CBC News. (2018, 18 April). Pedestrian bridge over new Turcot Interchange will be built after all. http://www.cbc.ca/news/canada/montreal/dalle-parc-pedestrian-bridge-turcot-1.4624772

Centre for Oral History and Digital Storytelling. (2013). *Walking the post-industrial Lachine Canal.* Concordia University.

Chapman, M. (2018). *Montreal climate coalition: For a carboneutral city.* Montreal Climate Coalition.

City of Montreal. (2018a, 16 April). The city unveils its vision of an international level, forward-looking downtown core. http://ville.montreal.qc.ca/portal/page?_pageid=5977,43117560&_dad=portal&_schema=PORTAL&id=29969&ret=http://ville.montreal.qc.ca/pls/portal/url/page/prt_vdm_en/rep_annonces_ville/rep_communiques/communiques

City of Montreal. (2018b). Unveiling of the new Rue Sainte-Catherine concept. https://www.makingmtl.ca/saintecath/news_feed/unveiling-of-the-new-rue-sainte-catherine-concept (password required)

Conseil régional de l'environnement Montreal. (2017a, 12 June). Dalle-parc Turcot: La mobilité durable restera un concept creux si le Gouvernement n'offre pas les infrastructures adéquates. News release.

Conseil régional de l'environnement Montreal. (2017b). Exigeons le retour de la dalle-parc!

Conseil régional de l'environnement Montreal. (2017c). Rassemblement pour la dalle-parc Turcot.

CTV News. (2010, 10 November). $3B Turcot plans draw detractors. https://montreal.ctvnews.ca/3b-turcot-plans-draw-detractors-1.572509

CTV News. (2018, 3 January). Major makeover on Ste. Catherine St. begins. https://montreal.ctvnews.ca/major-makeover-on-ste-catherine-st-begins-1.3743709

D'Amours, M. (2017, 11 June). Citizens renew push for pedestrian bridge connecting NDG to Southwest borough. CBC News. http://www.cbc.ca/news/canada/montreal/dalle-parc-project-transport-quebec-1.4156005

De Rosa, N. (2017, 20 April). Dalle-parc Turcot: l'objectif de pétition atteint, la mobilisation ne cessera point. *Métro.* http://journalmetro.com/local/sud-ouest/actualites/1121515/dalle-parc-turcot-lobjectif-de-petition-atteint-la-mobilisation-ne-cessera-point/

Developpement durable, E. e. L. c. l. c. c. (2018). Environmental Assessments. Gouvernement du Québec.

Fischler, R. (2009). What sort of problem is the replanning of the Turcot interchange? In P. Gauthier, J. Jaeger, & J. Prince (Eds.), *Montreal at the Crossroads: Superhighways, the Turcot, and the Environment* (pp. 79–90). Black Rose Books.

Gauthier, P., Prince, J., & Jaeger, J. A. G. (2009). *Montréal at the crossroads: Superhighways, the Turcot and the environment.* Black Rose Books.

Gould, Kevin. (2017, 11 April). Turcot consortium confirms it won't build Turcot bike route. http://montreal.ctvnews.ca/turcot-consortium-confirms-it-won-t-build-turcot-bike-route-1.3364991

Gravel, F. (2016, 4 December). Échangeur Turcot: la dalle-parc a disparu des plans! *La Presse.*

Gravel, F. (2018). TURCOT: les montréalais seront consultés sur une vision large des quartiers. *Turcot: Mobilisons-nous! Pour le retour de la dalle-parc.*

Gyulai, L. (2018, 30 March). Police wiretapped ex-mayor Gérald Tremblay, search warrants reveal. *Montreal Gazette.*

Harvey, D. (2013). Rebel cities. *Verso.*

Hodgson, F. C., & Turner, J. (2003). Participation not consumption: The need for new participatory practices to address transport and social exclusion. *Transport Policy, 10*(4), 265–272. https://doi.org/10.1016/j.tranpol.2003.08.001

Jones, P., & Lucas, K. (2012). The social consequences of transport decision-making: Clarifying concepts, synthesising knowledge and assessing implications. *Journal of Transport Geography, 21,* 4–16. https://doi.org/10.1016/j.jtrangeo.2012.01.012

La Communauté Milton Parc. (2002). Historique de l'échangeur Pins/Parc. https://www.miltonparc.org/historique-de-lechangeur-pinsparc/

Leavitt, S. (2016, 30 September). De la Concorde overpass: Before and after the collapse. CBC News.

Lefebvre, H. (1991). *The production of space.* Blackwell.

Lefler, D. E., & Gabler, H. C. (2004). The fatality and injury risk of light truck impacts with pedestrians in the United States. *Accident Analysis & Prevention, 36*(2), 295–304. https://doi.org/10.1016/S0001-4575(03)00007-1

Le Sud-Ouest Montréal, & Transports, M. d. e. E. d. T. Q. (2018). Plan de développement urbain, économique et social – Turcot.

Manaugh, K., & El-Geneidy, A. (2012). Validating walkability indices: How do different households respond to the walkability of their neighbourhood? *Transportation Research Part D, 16,* 309–315. http://doi.org/10.1016/j.trd.2011.01.009.

Martens, K. (2016). *Transport justice: Designing fair transportation systems.* Routledge. https://doi.org/10.4324/9781315746852

McAndrews, C., & Marcus, J. (2015). The politics of collective public participation in transportation decision-making. *Transportation Research Part A, 78,* 537–550. https://doi.org/10.1016/j.tra.2015.06.014

Mobilisation Turcot. (2011). The Mobilization.

Mobilisation Turcot. (2013a). Le projet Turcot reste toujours trop gros, sans transport collectif structurant et hors de contrôle.

Mobilisation Turcot. (2013b). Opération échangeur TRICOT.

Mobilisation Turcot & le Groupe de recherche urbain Hochelaga-Maisonneuve (GRUHM). (2012). Cure minceur pour l'échange Turcot.

Mobilisons-Nous! (2017a). Pétition.

Mobilisons-Nous! (2017b). Rassemblement.

Monno, V., & Khakee, A. (2012). Tokenism or political activism? Some reflections on participatory planning. *International Planning Studies, 17*(1), 85–101. https://doi.org/10.1080/13563475.2011.638181

Morency, P., Gauvin, L., Plante, C., Fournier, M., & Morency, C. (2012). Neighborhood social inequalities in road traffic injuries. *American Journal of Public Health, 102*(6), 1112–1119. https://doi.org/10.2105%2FAJPH.2011.300528

Mumford, L. (1964). *The highway and the city.* Mentor Books.

Nadeau, B. V. (2017, 11 June). Dalle-parc Turcot: les partisans ne baissent pas les bras. *Journal Métro.*

Norton, P. D. (2007). Street rivals: Jaywalking and the invention of the motor age street. *Technology and Culture, 48*(2), 331–359.

Pereira, R. H. M., Schwanen, T., & Banister, D. (2017). Distributive justice and equity in transportation. *Transport Reviews, 37*(2), 170–191. https://doi.org/10.1080/01441647.2016.1257660

Plourde-Archer, L. (2013). Montreal's Ruelles Vertes: Green alleyways help the environment and create a sense of community. *Untapped New York.* https://untappedcities.com/2013/08/07/montreals-ruelles-vertes-green-alleyways-help-the-environment-and-create-a-sense-of-community

Poitras, C. (2011). Montreal on the move: The surprising consequence of highways. In S. Castonguay & M. Dagenais (Eds.), *Metropolitan natures: Urban environmental histories of Montreal* (pp. 168–183). University of Pittsburgh Press.

Projet Montreal. (2018). Le Programme.

Public Health Agency of Canada. (2014). *Chief Public Health Officer's report on the state of public health in Canada, 2014: Public health in the future.* http://publichealth.gc.ca/CPHOReport

Radio Canada. (2013, 25 March). Québec révise le projet de l'échangeur Turcot.

Robert, M. (2014, 16 October). Chronique Montréalite no. 14 brève histoire des ruelles de Montréal. https://actu.fondationlionelgroulx.org/Chronique-Montrealite-no-14-Breve-histoire-des-ruelles-de-Montreal.html

Rodgers, D. (2012). Haussmannization in the tropics: Abject urbanism and infrastructural violence in Nicaragua. *Ethnography, 13*(4), 413–438. https://doi.org/10.1177%2F1466138111435740

Rodgers, D., & O'Neill, B. (2012). Infrastructural violence: Introduction to the special issue. *Ethnography, 13*(4), 401–412. http://doi.org/10.1177/1466138111435738

Scott, M. (2018, 20 February). Plante stands firm on closing Mount Royal to through traffic. *Montreal Gazette.*

Shaffer, M.-E. (2017, 26 July). La Charte du piéton, 10 ans après son adoption. *Métro.* http://journalmetro.com/actualites/montreal/1174444/la-charte-du-pieton-dix-ans-apres-son-adoption/

Sheller, M., & Urry, J. (2000). The city and the car. *International Journal of Urban and Regional Research, 24*(4), 737–757. https://doi.org/10.1111/1468-2427.00276

Shill, G. H. (2019). Should law subsidize driving? SSRN. https://ssrn.com/abstract=3345366 or http://doi.org/10.2139/ssrn.3345366

Soja, E. (2010). *Seeking spatial justice.* University of Minnesota Press.

Southworth, M. (1997). Walkable suburbs?: An evaluation of neotraditional communities at the urban edge. *Journal of the American Planning Association, 63*(1), 28–44. https://doi.org/10.1080/01944369708975722

Southworth, M. (2005). Designing the walkable city. *Journal of Urban Planning and Development, 131*(4), 246–257. https://doi.org/10.1061/(ASCE)0733-9488(2005)131:4(246)

Statistics Canada. (2006). Human activity and the environment: Annual statistics. https://www150.statcan.gc.ca/n1/pub/16-201-x/2006000/9515-eng.htm

Symon, J. (2017). New bike bridge over Turcot a go. *Montreal Times.*

Transport Canada. (2011). Active transportation in Canada: A resource and planning guide. Catalogue no. T22-201/2011E-PDF.

Transport Canada. (2015). *Canadian motor vehicle traffic collision statistics.* https://tc.canada.ca/en/road-transportation/statistics-data/canadian-motor-vehicle-traffic-collision-statistics-2015

Transport Quebec. (2009). Turcot complex reconstruction project. Le Bureau d'audiences publiques sur l'environnement.

TRICOT. (2013). Tricot pour la paix: Tricoter est une occupation pacifique. Solidarité Saint-Henri.

Twells, L. K., Gregory, D. M., Reddigan, J., & Midodzi, W. K. (2014). Current and predicted prevalence of obesity in Canada: A trend analysis. *CMAJ Open, 2*(1), E18–E26. https://doi.org/10.9778/cmajo.20130016

Ville de Montréal. (2015). Guide d'aménagement d'une ruelle verte 2015–2016. Le Plateau-Mont Royal.

Weaver Crawford, A. A. (2015). Reclaiming the ruelle: Creating a methodology for architectural interventions in the laneways of Montréal's Plateau-Mont-Royal. [Master's thesis]. Dalhousie University.

Wise, D. (2004). The physical design of alleys: An evaluation of alleys in the Montreal Plateau. [Master's thesis]. McGill University.

Yiannakoulias, N., & Scott, D. (2013). The effects of local and non-local traffic on child pedestrian safety: A spatial displacement of risk. *Social Science and Medicine, 80,* 96–104. https://doi.org/10.1016/j.socscimed.2012.12.003

3 Boroughs, Small Municipalities, and Sustainability: What Is Municipal Innovation and Can It Make a Difference?

RICHARD SHEARMUR

Introduction

Municipalities come in many shapes and sizes. When issues of sustainability are raised, metropolitan areas and mega-cities often come to mind (Seto et al., 2017), since environmental and social issues are most visible there. Networks of local governments formed to address sustainability primarily comprise medium-sized to large cities: the Urban Sustainability Directors Network includes 200 members across North America. Within Canada, it is composed of six major metropolitan cities and six large suburban municipalities. Yet in Quebec alone there are over 1,000 municipalities. All but six have populations of fewer than 150,000, and 930 of them have populations fewer than 15,000.[1] Furthermore, although large cities such as Montreal and Paris have central municipal governments, considerable powers and responsibilities are devolved to boroughs or arrondissements: these smaller municipal units, often with their own mayor and administration, tackle day-to-day problems, even as the central municipality takes positions and lobbies on a wider stage.

This chapter examines how innovation occurs in small municipal units, which comprise the majority of municipalities, as well as sub-municipal units (such as boroughs), which have an elected council and staff, and are in direct contact with citizens. These small units, often overlooked as actors in the quest for sustainable economic, social, and environmental futures, can play an important role. Not only can the local be a test bed for innovative ideas and processes (Bailly & Coulbaut-Lazzarini, 2018), small units can often innovate more radically than larger ones because they are out of the limelight (Grabher, 2018): provided that the local

1 http://www.stat.gouv.qc.ca/statistiques/population-demographie/structure/mun_15000.htm

population, officials, and politicians understand and agree to changes, local experiments with regulation, design, and participation can often outpace those of larger urban and metropolitan units. Although municipalities and boroughs also innovate in collaboration with private partners (raising questions of contractual arrangements, division of risks and rewards, intellectual property, etc.), in this chapter I focus on innovations developed within and by municipal and borough actors themselves.

The word "innovation" has been used to such an extent that it is difficult to know what it means (O'Bryan, 2013; Dwyer, 2014). In popular culture, as well as much academic literature, innovation is associated with new technology (such as iPhones, driver-less cars, robots, the internet of things), heroic entrepreneurs, and economic growth (Borins & Herst, 2018). Furthermore, since the end of the post-war golden years (i.e., since the oil crisis of the early 1970s, the rise of competition from developing countries, and the 1980s collapse of manufacturing in the West – see Fourastié, 1979), economic theory has taught us that GDP growth rests upon innovation (Solow, 1994). Why? Because, without market power, easy access to resources, and more capital or labour, the only ways to outgrow competitors are to produce goods and services of higher quality, to produce the same goods or services in better ways, or to introduce new products and services. All require innovation.

The notion that innovation is the driving force of the economy – maybe even of well-being – rests partly on observation and common sense, and partly on market-based and consumerist ideology. Observation and common sense tell us that achieving a goal with limited means requires inventing new practices and then implementing them. Likewise, if a new problem emerges (such as a drought or an unknown illness), old approaches and methods will not apply, and new ones will be necessary. Furthermore, humans are naturally inquisitive and tend to delight in novelty for its own sake, though these inclinations are tempered by the need for identity and stability. Hence there is cultural tension between innovation and tradition (Schwartz, 2012). Innovation to further a common goal – whether GDP, sustainability, or mass murder – is inherently normative: the nature of the goal, the trade-offs required, the value in achieving the goal – require debate and judgment. Innovation for its own sake may be stimulating but it is not inherently desirable, nor does it inevitably lead to increased well-being, however defined.

This normative aspect of innovation explains why it has become an ideological issue (Wagenaar & Wood, 2018). The goal to which economics and most government policy have harnessed it, and to which it is assumed that society agrees, is economic growth (Fioramonti, 2014). The natural inquisitiveness of people is understood as a means to further consumption

(Baudrillard, 1970), and the tension between innovation and tradition has been resolved in favour of innovation. Thus, whilst innovation is natural (and sometimes worthwhile), and whilst resistance to innovation is also natural (and sometimes worthwhile), since the 1970s – and maybe since the beginning of the Industrial Revolution – emphasis has been placed on the *qualities* of innovation and the *evils* of resisting it. The term "innovation" carries positive normative baggage: it is considered "good," and questioning it is disparaged as Luddite (Elliott, 2014).

Within this context, in this chapter I describe how innovation occurs in smaller municipalities and boroughs, and how they are positioned to identify and develop small-scale solutions to sustainability. That these solutions are small scale reflects the size of municipalities and boroughs and their particularities. However, small scale does not mean insignificant – solutions and approaches can be shared and adapted: through networks of municipal actors, these solutions have potential for wider effects. This does not obviate the need for concerted policies and action in metropolitan areas, mega-cities, and nations at a global scale, but it opens avenues of action and experiment (i.e., of innovation) that can be implemented, shared, and improved upon locally, a scale immediately accessible to citizens. Given the prevailing view of innovation and its purpose, it is necessary to first theorize how municipalities can innovate (since their purpose is not to augment GDP nor to satisfy markets) and how their innovation can be evaluated, and to then examine some examples.

The chapter proceeds as follows. First, I discuss what municipalities are and why it is important to understand how they innovate. Governments, in particular local governments, are often singled out for their slowness and lack of entrepreneurial spirit (Mazuccatto, 2014; Shearmur & Poirier, 2016; Kastelle, 2015[2]). I suggest that their slowness (partly) reflects salutary features of government, and that innovation can occur within municipalities in a fashion similar to the private sector, but with key differences. Given the efforts of New Public Management to introduce market-like dynamics into the public sector (Hood, 1991), I carefully set out why a comparison with the private sector is useful, but also why this does not necessarily entail competition, the profit motive, or markets (Swann, 2014).

---

2 This perception is held by many proponents of New Public Management – who argue that public administrations should be run as businesses so they can become equally nimble, efficient, and innovative (see Hood, 1991) – and is reflected in public perception, even though academics have long recognized public-sector innovation: Kastelle (2015, p. 68) reports that "one question that I often get asked is 'can the public sector innovate at all?'"

I then briefly describe some examples of innovation, most of which have been presented at Quebec's annual Municipal Innovation competition, run since 2005 by the Union des municipalités du Québec. These innovations are connected, directly or indirectly, to sustainability. This connection is unsurprising, since municipalities and boroughs are the level of government that manages day-to-day material flows of waste, traffic, and people, and oversees the infrastructure that provides clean water, safe roadways, and the delivery of energy. I conclude that municipal innovation can have more impact than its small scale suggests. Following Gibson-Graham's (1996) arguments about local and small-scale action, municipal innovation not only empowers local citizens and politicians by addressing important local environmental issues, it can also have wider effects by virtue of the networking and sharing that occurs locally, nationally, and internationally among municipal actors.

Municipal Innovation: A Schumpeterian Approach

In this section I describe key features of smaller municipalities and boroughs, which I approach as organizations. I then describe the Schumpeterian innovation process, which permeates much economic and management literature on the topic (Scherer, 1986), and argue that it applies to innovation of all sorts (not merely to market-driven innovation), provided that some assumptions are modified. These assumptions concern the motivation for and the purpose of innovation, not its basic process. To conclude this section I return to municipalities and boroughs, outlining how the Schumpeterian approach can help understand innovation within these organizations.

What Is a Municipality?

Municipalities (or boroughs within large municipalities) are the smallest and most local unit of government in most countries. In Canada, for example, the constitution divides power between federal and provincial jurisdictions and does not mention municipalities. They are "creatures of the province," i.e., local entities created by provinces, often at the behest of people wishing to self-organize locally, to which certain provincial powers are devolved (Magnusson, 2005). These powers generally cover local sanitation, waste disposal, water treatment, urban and environmental planning, local roadways, and provision of local services (which can extend to schooling, day care, sports, libraries, housing, etc.). Municipalities can enact by-laws and have powers to enforce them, but, in Canada at least, the province can almost always override municipal by-laws and decisions, albeit following a codified procedure. In France, communes and arrondissements, similar to municipalities and boroughs, are more autonomous, since they are enshrined in the constitution (Serieys, 2018). However, the French state is centralized and in most cases can override or bypass communal by-laws and decisions. Given their rather tenuous power relationship with higher levels of government, municipalities survive and prosper by providing necessary services to their citizens, and – being close to them – by drawing political blood should upper levels of government over-exert their powers.

A municipality or borough is not a straightforward entity. It is composed of numerous "parts" and can be apprehended from a variety of angles. Whilst the previous paragraph focuses on their constitutional and legal character, municipalities are also territories, i.e., delimited portions of inhabited land over which a local government has some power and for which it bears some responsibility. Local government is embedded in territory, since it is usually composed of a mayor and local council elected by residents and usually expected to be residents themselves. By virtue of this closeness to residents and land, municipalities or boroughs[3] are the level of government most in tune with day-to-day opportunities and challenges that present themselves in the territory. Thus, a municipality is not only a legal entity with power over geographic terrain; it also comprises citizens – both residents and organizations – that operate there. At the municipal level many consultations take place, and it is the level of government at which it is easiest (but not always easy!) for citizens to dialogue directly with politicians and administrators (Wills, 2016).

Municipalities are also organizations, and they will be approached as such in the rest of this chapter. The municipal organization comprises a permanent administration and departments, which organize and manage the municipality's finances, services, road repairs, libraries, and other municipal tasks. In many ways, but not in all, a municipality resembles a firm. It has a municipal (corporate) identity, it has a mayor (director), council (board), administration, and departments. It also comprises citizens (clients) with whom interactions are necessary and from whom feedback is important. It is circumscribed by strict legal boundaries, but has some leeway within them.

Municipalities differ from firms in important ways. There are no shareholders – a municipality is more like a cooperative, since residents elect the mayor and councillors, and residents are its main beneficiaries. A municipality does not operate in a straightforward competitive environment: notwithstanding Tiebout's (1956) economic argument that residents shift

---

3 I will use the term "municipality" to denote municipalities or boroughs and only refer to boroughs when I am not also referring to small municipalities.

municipalities in search of their desired balance between taxes and services, the causal direction of the relationship is not clear (Howell-Moroney, 2008). A social interpretation of the residential mobility observed by Tiebout suggests a clustering hypothesis, i.e., that residential choice is driven by status considerations: having clustered, residents of similar status will then vote for administrations that ensure their desired balance of services and taxes. Gentrification is a good example: precursor residents are attracted to a neighbourhood because of its buildings, diversity, or other features; these precursors make the neighbourhood desirable, leading to gentrification. In the process, governance structures (taxes, schools, neighbourhood watch, residents' associations, planning, etc.) are adapted to gentrifiers' wishes (Tissot, 2015). There is, of course, pressure upon municipal administrations to deliver good services, to identify and solve local problems, and to do so without increasing local taxes unduly. However, this pressure arises not principally through competition, but internally from citizens.

Municipalities face a complex set of objectives, rendered more so because "good" service, "undue" increase, and "problem" mean different things to different people. Whereas a company's bottom line is usually profit – and to some extent market share – in view of satisfying shareholders (and maybe directors), the goals and beneficiaries of municipal action are multiple, intertwined, and often difficult to reconcile. Consultative, discursive, and political processes are therefore necessary.

A Schumpeterian Approach to Innovation

Joseph Schumpeter, an Austrian economist working in the first half of the twentieth century, was amongst the first to clearly articulate innovation as an economic process (Schumpeter, 1936; Scherer, 1986). As usually understood – that is, assuming innovation responds to market demand – this process is broken down into three main elements, connected by a fourth.

1 *An idea or concept for a new or improved product or service.* This can come from formal scientific research, creative insight, or transposing ideas from one field to another.
2 *Idea or concept development.* This requires resources to implement initial versions of the idea, and also requires testing, such as in labs, focus groups, or pilot projects.
3 *A market or some other type of demand.* For a prototype to become a successful innovation, there must be demand. Even if a concept can be transformed into a workable prototype, the innovation will not necessarily be successful. An innovation moves beyond the prototype stage when it becomes commercially viable and begins either to replace

previous similar products or services, or to generate new demand for previously non-existent products or services. In both cases older technologies, products, and services will tend to disappear – hence the term "creative destruction."[4]

4 *The key actor who draws these items together: the entrepreneur.* This person or company requires vision, the ability to marshal resources, and the capacity to market the innovation. An idea or concept remains an idea or concept if nothing is done about it. The entrepreneur, motivated by profit, identifies the potential, understands the demand, convinces backers to finance development and testing, and organizes the marketing.

From the 1980s onward it became evident that the innovation process is more complex, incorporating feedback loops (Chesbrough, 2003). Concepts and ideas are no longer conceptualized as "being out there" – rather, entrepreneurial firms develop them by interacting with a wide variety of external actors (clients, collaborators, competitors) in a variety of contexts (fairs, meetings, formal collaborations, social events, internet). Concepts and ideas evolve as clients, partners, and the entrepreneur gain insight into the product or service and its possible uses.

In the next sections I show how the Schumpeterian innovation process is a useful framework for understanding non-market innovation, and how, in particular, it provides a way of understanding how municipalities (as organizations) innovate in areas related to sustainability.

How Does Schumpeterian Innovation Relate to Non-Market Innovators?

Recent advances in economics and management studies have deepened and extended Schumpeter's theory of innovation without questioning the basic processes and actors it outlines.[5] However, Schumpeter's ideas

---

4 The idea of creative destruction comes from Marx, who introduced the concept to describe how capitalism destroys previous economic and social orders. However, the term is most often used today with reference to Schumpeter and to the role and effects of innovation.

5 The discipline of evolutionary economics has built extensively upon Schumpeter's work (Andersen, 1994), and some of his conclusions – such as those outlined in his *Capitalism, Socialism and Democracy* (1942), which are macro-economic – are much debated today. My comment about his ideas being broadly accepted relates specifically to his micro-economic discussion of the entrepreneurial process, and to the process of creative destruction (abstracting, though, from resource concentration and from advantages that incumbents have over innovators: these underpin his later ideas, which are particularly relevant today as the platform economy generates huge and apparently unassailable monopolies; see Srnicek, 2016).

rely on markets, entrepreneurs motivated by profit, and risk-taking: the market is the final arbiter of whether a prototype becomes an innovation, and the profit that motivates entrepreneurs is justified because they take risks.

To what extent, then, can Schumpeter's ideas shed light on innovation in non-market environments – in civil society, community organizations, or municipal organizations? A great deal. In the subsections below I first generalize the four steps described above so that they also apply to non-market innovation. I then consider how innovation can occur within non-market organizations, whether incremental or imitative changes can be considered "innovative," and how the public sector's duty of care alters the way innovations are introduced.

THE NON-MARKET INNOVATION PROCESS

Non-market innovation (social, community, or municipal) requires an idea or concept. The concept needs to be developed and tested, and it must successfully respond to some form of demand (i.e., it must be successfully implemented). However, in a non-market innovation, demand is not expressed in customers' willingness to pay (although it may be): demand is expressed in different ways, such as a problem that needs solving (e.g., ensuring that potholes are identified and filled), a social need that should be met (e.g., taking meals to older people), or an agreed-upon aspiration that should be fulfilled (e.g., reducing fuel use).

Likewise, the value created by innovation need not be monetary: it can be environmental, aesthetic, ethical – these values are not reducible to money (Anderson, 1995).

Such alternative ways of understanding value require consensus on goals – on the innovation's purpose – and on acceptable resources (time, materials, money) to be invested. Non-market innovation does not require willingness to pay but does require agreement that it addresses a problem, fulfils a need, or gets society closer to realizing an aspiration. "Agreement" need not be unanimous, but for an innovation to be deemed successful, agreement is required amongst stakeholders: agreement should not only cover whether the innovation successfully addresses the issue, but whether it does so in an appropriate and resource-effective manner. For market innovations, it is market processes that synthesize this information and produce market-generated agreement. In non-market situations, agreement emerges from discussion and from comparing the innovation's outcomes with agreed-upon goals and criteria: it is discursive and political (Wills, 2016).

Given these important provisos, the Schumpeterian innovation process can be restated in more general form, extending it to non-market innovations. An innovation, whether market or non-market, requires:

1 *An idea or concept for a new or improved product, service, process, regulation, or way of doing.* This can come from formal scientific research, creative insight, discussion groups, politicians, or citizens, or from transposing ideas from one field to another.
2 *Idea or concept development.* This requires resources to implement initial versions of the idea. It also requires testing, such as in labs, focus groups, or pilot projects. At this stage the nature of the value being created and/or the goals of the innovation will be clarified and, if this is a non-market innovation, will be subject to discussion and negotiation.
3 *Successful implementation and adoption.* For a prototype or pilot project to become a successful innovation, it must be implemented. And the community, its users, and/or its promoters must recognize that it has attained the goal (i.e., created the value) that was expected and has efficiently used resources. Previous products, services, processes, regulations, or ways of doing will tend to disappear.
4 *The key actor who draws these items together: the entrepreneur.* This person, company, or organization requires vision, the ability to marshal resources, the facility to explain the innovation and clarify its purpose, and the capacity to develop and implement the innovation.

ENTREPRENEURSHIP IN MARKET AND NON-MARKET ORGANIZATIONS

The problematic dimension of the Schumpeterian innovation process, when applied to non-market entities such as municipalities, is the entrepreneur. In a market context this person or firm is a risk-taker, who not only understands the potential for innovation but gathers resources to develop the idea into a prototype, test it, and market it. He or she is motivated by profit, i.e., remuneration that is conditional upon successful risk taking.

Although entrepreneurial firms are usually understood as small firms, innovation also occurs in larger firms, for which the idea of entrepreneurship is also valid (Casson, 2005), and that have some similarity (in organizational structure) to municipalities. Within larger firms, it is often departments or individuals who identify a potentially innovative idea or concept, convince their supervisors or directors that it is worthwhile, and marshal resources to develop it. This intra-firm entrepreneur is a risk-taker: just like market entrepreneurs (Shane & Cable, 2002), intra-firm entrepreneurs seldom risk their own resources: rather, they stake their reputation with their backers. Finally, within a large firm the entrepreneur is not motivated by sizable personal gain: there may be a bonus or reward if the innovation is successful, but most profits will devolve to the firm. The motivation of intra-firm entrepreneurs may be inquisitiveness, interest in solving a particular problem, or longer-term career advancement.

It is a small step from describing intra-firm entrepreneurship to arguing that entrepreneurship occurs within municipalities and other non-market organizations. Entrepreneurs within municipalities can be politicians but can also be members of staff, administrators, or even engaged citizens. Typically,[6] within a municipality or borough, the instigator of an innovation – that is, the entrepreneur – will be a staff member or the head of a specific department (such as the head librarian, the director of the computer department, the director of arts and culture) (Borins, 2014). This person will have identified a problem that can potentially be addressed, or will have come up with a new idea, such as a new service that can be offered or a new way of performing an activity. Municipal entrepreneurs are rarely motivated by financial gain, since the goal of the municipality (and of most people making a career in such organizations) is not profit. The motivation is usually inquisitiveness, public service, or feelings of civic duty (Perry & Hondeghem, 2008), though career advancement and other personal motives may play a secondary role – secondary because reputational risk usually outweighs foreseeable personal reward (Casebourne, 2014).

Depending on the scale of the idea, it will either be developed within the municipal department (with resources allocated by the department head), or at a wider scale with budgets allocated by the mayor and councillors. In both cases, the municipal entrepreneur will have put his or her reputation on the line and have demonstrated an ability to convince backers. In all cases, elected politicians and high-level administrators play an important role: whether or not they are the source of the initial idea, whether or not they identify an area in which innovation is required,

---

6 Much of the discussion in this chapter is based upon the author's experience as judge on the MériteOvation innovation competition run by the Union des municipalités du Québec, as well as his extensive work on innovation in small and medium-sized enterprises (SMEs). Since studying the competition from a research perspective in 2015 (see Shearmur & Poirier, 2017), he has sat on the jury in 2016, 2017, 2018, and 2019. This has not only allowed him to closely examine particular examples of municipal innovation but has also provided the opportunity to discuss with municipal innovators themselves and with other jury members the processes that lead to innovation. The word "typically" is used here in reference to this experience.

and whether or not they are directly involved in approving development budgets, they create an environment within which entrepreneurial risk-taking is either allowed (or indeed encouraged) or disapproved of (Borins, 2014). It is only if elected politicians and their administration are open to new ideas and to a certain degree of risk-taking that entrepreneurship can occur within a municipality.

There will typically be a period of research and development, exploring similar projects in other municipalities, examining academic research, experimenting, or bringing in consultants. In certain cases of small-scale innovation – usually developed to solve a specific problem faced by employees within their daily activities – development is carried out internally almost exclusively. Once a solution is found and tested, it is piloted, and, if successful, deployed within the municipality. It is at this stage that the concept / invention / prototype can be called an innovation.

IS NON-MARKET MUNICIPAL INNOVATION INNOVATIVE?

The economic approach to innovation implies that firms introduce innovations that are sufficiently original that they do not infringe copyrights or trademarks and that allow the firm to out-compete rivals. This is not so in municipal innovation: municipal entrepreneurs will often look towards other municipalities to examine possible solutions – municipal innovation is often imitative or incremental. Furthermore, municipal innovations are rarely patented (most *cannot* be, since they relate to processes and services, not to products and technologies), so there is no formal way of ascertaining whether a municipal innovation is a first or whether it mimics something that has been introduced elsewhere.

The question of originality speaks to a misunderstanding, or maybe a shift, in the meaning of "innovation." As Godin (2008) documents, until the 1960s, "innovation" was often a synonym of "imitation": a firm or an organization innovated if it introduced a product or process new to the firm or organization, even if it already existed elsewhere. This understanding still prevails within the community that studies innovation: the OECD (2005), as well as most other students of innovation, recognizes that some innovations are "new to the firm," others are "new to the local market," and others are "world firsts." However, the understanding of innovation that prevails more widely is of world-first or ground-breaking change – to such an extent that minor changes in the functionality of products such as iPhones are marketed as major innovations since otherwise they would not be not valued (Borins & Herst, 2018).

Hence, the question of absolute novelty is moot: a municipality innovates if it successfully implements a product, process, or technique that is new to it. The innovation is successful if it meets agreed-upon goals and resource requirements. The novelty may apparently be absolute (i.e., no precedents have been found), or relative (i.e., precedents have been adapted to suit the requirements of the particular municipality).

INNOVATION AND ITS CONSEQUENCES: MUNICIPALITY'S DUTY OF CARE

There is a difference between innovation in the municipal and public sphere, and innovation that is carried out in firms or by small civic associations. Municipalities have a duty of responsibility towards citizens, and rolling out an innovation can affect their lives, including the lives of citizens who do not want the innovation.

For instance, an innovative method to treat waste-water may have been developed by a municipality, but the consequences of its failure if it is deployed are catastrophic and widespread. Therefore, a municipality has a duty of care, and, indeed, of conservatism, which is not present to so great a degree in the private or voluntary sectors (Potts, 2009). At a more modest scale, changing the way waste is collected (for example) can upset many habits and routines for local people and businesses who may not want the disruption: even apparently minor innovations of this sort need to be implemented by municipalities with their ramifications and consequences in mind.

All innovators are expected to respect regulations and standards designed to limit negative externalities (such as pollution, noise) and to establish a predictable framework within which social and economic activities can occur. However, regulations can hold back and limit innovation: some innovators, such as Uber and Airbnb, argue that regulations should therefore not apply. This is rational (for them) because they will not bear the external costs of their innovation, whereas they reap its benefits: external costs will be borne by cities, incumbent businesses, and local residents (Bell, 2016; West & Lu, 2009; Swann, 2014). A municipality – although it has the formal capacity to alter certain regulations and standards, whilst being constrained by those set by higher orders of government – could not force innovation on its citizens in this way without paying a political price: its duty of care, combined with the fact it represents the people who will benefit *and* suffer from the consequences of innovation, means that it could not introduce an innovation (including a new regulation – Reich & Schleicher, 2015) without carefully considering all consequences, positive and negative. This can take time and can slow the rate of adoption of innovations.

Municipal Innovation and Intellectual Property, Big Data, and Technology Lock-In

So far I have discussed the process by which an individual, organization, or department identifies the potential of an innovation and carries it through. There is another understanding of municipal innovation: the purchase and implementation of technologies or services provided by private firms. In the 1980s, for example, service innovation was understood primarily as the purchase and implementation of computer technology (Barras, 1986), and during the 1990s and 2000s, under the influence of New Public Management, many public-sector agencies innovated by outsourcing services (Diefenbach, 2009; Wagenaar & Wood, 2018). This approach is taking on renewed relevance as "smart cities" gain momentum (Greenfield, 2013): many cities are now working with large corporations such as Google (Oliviera, 2018) or Philips (Meijer & Thaens, 2018) to develop ways of gathering and harnessing data about inhabitants, traffic flows, waste, energy use, etc., in order to manage them more efficiently.

This type of innovation is controversial for a variety of reasons. First, cities are handing over vast amounts of data to corporations, which can then use and market them for unspecified purposes (Morozov, 2017). Second, the "problems" that these data-driven innovations address have not been identified by municipalities: rather it is corporations that are in search of testing grounds and new (public) markets for their technologies (Shearmur, 2016). Finally, although municipalities participate in the development of these technologies, they obtain little in return: municipalities hand over the intellectual property they help create in return for first-generation implementations that will need updating as technologies develop.[7] Recent research on (innovative) subcontracting in the 1980s and 1990s reveals that service quality declined, work conditions of front-line staff deteriorated, whilst costs increased, principally because private providers were effectively handed monopolies (Hood & Dixon, 2015; Bowman et al., 2015).

An argument can therefore be made for municipalities to develop their own technologies, services, and solutions, and/or for them to cooperate in creating open-source platforms, sharing innovations they develop.

---

7 This argument, developed by Mazzucato (2015) for public-sector technologies, is alluded to in the context of urban technologies by Meijer and Thaens (2018) in their conclusion, and is discussed, again in an urban context, by Shearmur (2016).

There is a tradition of policy transfer between public bodies (Dolowitz & Marsh, 2000), municipalities in particular (Temenos & McCann, 2012; Harris & Moore, 2013[8]), which may disappear if municipalities no longer have the departments and internal resources necessary to develop, test, and implement innovations. The municipal innovations described in the next section, even if they sometimes call upon existing technology, are primarily developed, driven, and owned by municipalities themselves.

Municipal Innovation and Sustainability: Some Examples

So far, little has been said in this chapter about sustainability, except that, because municipalities and boroughs focus on green space, roads, traffic flow, waste management, and other local services, the nature of their activities places them at the forefront of issues relating to sustainability.

This section presents some examples of innovation in environmental sustainability. Like innovation, "sustainability" is a word that can encompass many things. Whilst I have theorized and described innovation (because innovation in municipal organizations, or indeed in other non-market settings, has rarely been conceptualized[9]), I will be more succinct regarding sustainability. Sustainability can be viewed as an agreed-upon goal associated with social, environmental, and ethical values – an overall objective that governs the actions and decisions of many – but not all – public bodies. From an environmental perspective it involves making decisions and implementing processes that tend towards reducing use of non-renewable resources, reducing pollution, and increasing equity in view of handing over the planet to the next generation in an ecological and social state as good as, if not a better than, it was. Many international bodies (UN, 2018), nations (SCESD, 2017), cities (Montréal, 2016; Temenos & McCann, 2012), and municipalities (Drummondville, 2018) enshrine this general principal into policy documents, though sustainability goals can be easy to ignore when pitted against others such as economic growth (Temenos & McCann, 2012; Sachs, 2018).

---

8 Note that Temenos and McCann (2012) refer to this as "policy mobility" and critique it for tending to disseminate prevailing neoliberal ideas. Harris and Moore (2013) show that policy sharing between municipalities has a long history, particularly in urban planning.

9 An exception, who not only describes but theorizes small-scale non-market innovation (though not in the context of municipal organizations) is Swann (2014). Swann sets out a tightly argued case for the existence and importance of innovation driven by individuals and communities for non-market purposes (often for community or environmental objectives).

Sustainability: Why Consider Municipalities and What Can They Achieve?

Municipalities and boroughs can act only at a small scale: however, their actions can have major impact collectively. For instance, in Quebec there are 185,000 kilometres of roads, 31,000 of which are provincially managed (highways and state roads), 92,000 municipally. The remainder are administered by ministries or para-governmental bodies (Québec, 2018). Thus, should municipalities experiment and establish, for instance, an ecologically friendly way of maintaining roads, it would have significant impact if rolled out across a majority of municipal roads. Likewise there are 305 public libraries in Quebec, distributed across 165 municipalities: should these libraries introduce new methods, new ways of engaging younger and older readers, this would have an impact on equity across the province. Whilst each municipality is small, collectively they are responsible for, and manage, vast resources. Furthermore, the large number of municipalities means there is place for natural experimentation – the successful ones can share their ideas.

The idea behind Quebec's MériteOvation municipal innovation competition is to provide a forum where successful local innovations are given publicity and where other municipalities can seek inspiration. However, as Wills (2016) points out, municipalities have different means at their disposal, and not all are able to implement new ideas, even if desirable: municipal innovation *could* have a major impact, but higher levels of government remain crucial for ensuring inter-municipal equity, and for supporting the propagation of good ideas.

It is with this in mind that I briefly describe some examples of municipal innovation that chip away at the global issue of sustainability.

Some Examples of Municipal Innovation That Relate to Issues of Sustainability[10]

Innovation can consist of major changes to the way things are done or can be small. In this section I point the reader to a variety of innovations that illustrate how the proximity of municipalities to everyday material and social problems can make them key actors in altering procedures and habits in view of increasing sustainability.

---

10 Municipalities also engage in sustainability practices linked to equity – providing library services, internet access, services to the elderly, social housing, and holiday camps for children. To save space, the following examples deal only with environmental sustainability.

BIO-METHANIZATION IN SAINT-HYACINTHE, QUEBEC[11]

In 2014 Saint-Hyacinthe, a town of about 50,000 in central Quebec, opened a bio-methanization facility, which generates usable gas from bio-degradable waste. Although a smaller bio-methane plant had already been introduced in Berthierville, Quebec, by a private energy company in 2003, this project was initiated by the town's engineers, one of whom had an interest in the process. Visits were made to existing plants, mainly in Europe (there were few, if any, in North America, and none of the scale envisaged in Quebec) to understand the technologies involved and prepare a strategy. In parallel, the city investigated provincial and federal grants for implementing waste recycling and greenhouse gas reduction. The plant represents a $48 million investment, which was shared by the three levels of government.

This can be understood as essentially imitative innovation: an existing technology, albeit one not deployed in North America at the scale envisaged, was studied, visits were made to gather information, and the technology was adapted and deployed in Drummondville. The impacts of such imitation should not be underestimated: the plant is now running successfully, generating methane from local agricultural waste, from local institutional waste, from school-lunch leftovers, and from residential and commercial waste. Waste is also being gathered from 25 other municipalities.

Whilst this imitative technological innovation is noteworthy, perhaps of more import is how the bio-methanization plant was integrated into a wider policy framework. As the plant opened, the city began converting its vehicles to methane gas and heating its buildings with the fuel – making it part of an overall strategy – with excess gas sold to Quebec's main gas supplier (this is itself a regulatory innovation, the first time Gaz-Métro purchased municipal bio-methane; Montpetit, 2018). The plant opening was accompanied by a concerted effort to inform and educate local businesses and school children, encourage the separation and collection of biodegradable waste, and raise awareness of the importance of recycling.

As Montpetit (2018) argues, even this innovation – which is of considerable size for a municipality – is small: at the scale of Quebec, let alone the globe, reducing CO_2 emissions by 49,000 tons annually is a minor achievement. However, the project not only illustrates (to other municipalities) that such endeavours are possible and profitable, it imitated

---

[11] Information on this can be found at UMQ (2021). It has been reported on extensively in the media.

similar projects in Europe: the sharing, diffusion, and local adaptation of municipal innovations are a key way they can achieve wider impact.

TERREBONNE'S ENVIRONMENTAL PATROL AND SALABERRY'S WATER[12]

Although environmental sustainability is now a widespread preoccupation, the concept – particularly its application at the municipal level – was not widespread in the early 2000s. In 2005 the town of Terrebonne implemented a local environmental policy, including the collection of biodegradable waste, water-use restrictions, recycling, and sites for gathering hazardous waste (electronics, etc.). Such a policy, new to the town, had many counterparts elsewhere, even if it was fairly new to Quebec.

However, the essence of much municipal innovation is the way towns incrementally improve on existing ways-of-doing, and how, by virtue of their multidimensional role, they cross-fertilize innovations. In this case, the innovation is not the environmental policy itself, but the effort the town made to engage its young and older citizens in the process. Over the summer of 2005 the town established an "environmental patrol" of four young people. These people visited neighbourhoods, parks, and individual houses, explaining the policy, gathering feedback, and encouraging residents to respect the town's regulations. The engagement of young people, and their efforts to discuss the policy with all residents, allowed the new environmental rules to be implemented with relatively little friction: the values and purpose of the innovation were explained and disseminated, increasing the program's effectiveness.

Another example of cross-fertilization is the renovation of Sallaberry's water-treatment facility on the St. Charles River, when a new filtration plant was needed to respect tightened provincial regulations. There is nothing particularly innovative in updating a filtration plant, but the city found it an opportune time to landscape the area surrounding the works: the banks of the river were planned as a leisure park, making the river accessible for kayaking and other water activities. This is a good example of an opportunity seized by an entrepreneurial municipality, cross-fertilizing a necessary quasi-industrial intervention, with enhancement to the environment. Again, this can be interpreted from the perspective of value: the council recognized that other (environmental and amenity) values could be enhanced whilst furthering externally mandated goals.

---

12 Information on these innovations was reported in a 2006 issue of *URBA* (UMQ's quarterly magazine), consulted at UMQ's archives. Terrebonne's innovation was reported by Ladouceur (2006), Salaberry's in the magazine *Public Action* (2006).

Both innovations could be shrugged off as minor interventions in small towns. However, both were successful, both were implemented at low cost (the first for the cost of a few summer internships, the second by integrating environmental considerations at the design stage), and both provide examples to other municipalities of how environmental values can be integrated into municipal activity.

YAMASKA'S ATTEMPT TO MANAGE ITS WATERWAYS[13]

Water runs through municipalities. No matter how polluted the water may be, municipalities have limited capacity to influence it, since they are responsible only for what occurs on their territory. In Quebec, municipalities are grouped into *municipalités régionales de comté* (MRCs), which typically comprise about ten. Whilst certain planning competencies can be devolved upwards to the MRC if municipalities agree (such as regulating drainage, determining the distance of facilities to waterways, and fining contraveners), the problem of water flow is still critical: what can an MRC, which only ever comprises portions of water-sheds, achieve?

In 2017, in this context of regulatory sclerosis, the MRC of Yamaska formally introduced a water management plan, whose regulations limited phosphorus run-off into rivers and lakes, tightened the use and maintenance of septic tanks, and introduced inspection of all waterways and lakeshores to ensure provincial regulations were applied. During the first year (2017) inspectors handed out notices, but since 2018, fines have been levied. The MRC is also conducting a wetland survey in view of protecting them in the same way.

This regulatory approach is innovative because it has overcome conventional wisdom that little or nothing can be done locally about water pollution (Kahn et al., 2015) and has given some (local) teeth to provincial regulations that remain essentially toothless for lack of enforcement. The plan, whilst run and implemented by the MRC, required agreement and cooperation from all municipalities that constitute it, as well as from citizens and lobbies (such as the agricultural lobby) whose members must respect the regulations.

This innovation does not overcome the limited impact of local intervention on water pollution. However, it *has* demonstrated that local authorities are not powerless. Furthermore, Yamaska MRC is now negotiating with other MRCs of the Yamaska watershed to widen the regulatory initiative. Furthermore, municipalities and MRCs of other

---

13 See L'OBV Yamaska, http://www.obv-yamaska.qc.ca/.

watersheds have shown interest in this initiative, which has powerful demonstration value.

EFFICIENCY GAINS, INTELLECTUAL PROPERTY, AND COMPUTER SOFTWARE

A growing number of municipalities are developing computer software in-house, for two reasons. First, commercial software is often ill-adapted to municipal requirements. Municipalities operate under strict oversight from upper levels of government, from their own taxpayers, and from bodies that regulate standards. Reporting requirements are specific to each jurisdiction. Furthermore, within these constraints – which reflect governments' duty of care for their citizens – municipalities are faced with context-specific problems and particular requests from citizens and elected officials. Off-the-shelf software is rarely well-adapted for such needs.

The second reason for developing software in-house is cost of ownership, updating, and adaptation. Software bought off-the-shelf and/or adapted or programmed by consultants is seldom owned by the municipality, while further changes or updates engender considerable cost. Even though developing software in-house is perhaps a slower process, it ensures the municipality has internal capacity to update and alter it, as well as the legal right to do so. As internal capacity develops, municipalities become better at developing their own software, gaining independence from licensed products.

Whilst internally developed software can serve many purposes, it has been used to increase the efficiency of municipal services, thereby reducing resource consumption. Trois-Rivières, Quebec, has been at the forefront of this area. In the early 2010s the city took upon itself to increase snow-removal efficiency (UMQ, 2021b). In countries such as Canada it is a major operation, involving multi-million-dollar annual budgets, fleets of heavy trucks and snow-removal equipment, and disruption to residents. Building upon basic software such as ArcGis and GPS trackers available in retail stores, the city developed a system that optimizes the deployment of vehicles, tracking them in real-time and adjusting clearance operations when necessary. It also keeps citizens informed, via an app, of snow removal plans and current operations. Not only have snow removal costs been reduced by about 15 per cent (mainly by reducing machine idling time), fuel consumption has also declined, saving an estimated 16 tonnes of CO_2 emissions (18 per cent) per year.

In 2018, the city was again recognized by the UMQ for a software innovation, a system that optimizes management of city contracts (UMQ, 2018). Its principal purpose is to centralize information, so that managers and elected officials have an overall and up-to-date understanding of the city's contracts, and of current and planned commitments. Information on all contracts and contractors is available, and their conformity with regulations and standards is ensured – partly by checklists when the information is entered, partly by daily cross-checking subcontractors with provincial and federal listings for fraud and non-compliance.

This system has considerable environmental impact because it allows each department and each contracting party to see not only which contracts are underway, but which are being discussed, which have been budgeted for, etc., over the next few years, and where, geographically, physical work will occur (if such work is involved). Thus, items that need attention (for instance, the replacement of an aqueduct) are flagged and geo-located, even if no contract has yet been signed. This has led to increased coordination among departments: for instance, if a road needs excavation for aqueducts, electricity, and refurbishment (three different departments), it is now easier to coordinate and minimize disruption and resource use.

These two systems, developed by the municipality, have been made available to other municipalities. Now the challenge for them is to build their own internal capacity to adapt and maintain the systems.

BOROUGHS DRIVING INNOVATION WITHIN LARGER CITIES

Montreal, like many large cities, has adopted a strategy to reduce greenhouse gas emissions. Many citizens perceive it as a necessary but top-down set of regulations. But some boroughs are innovating. One example is Rosemont-La-Petite-Patrie. In 2009 it had only four green alleys, which are back alleys turned into garden-like spaces, often with an urban agriculture component. By 2017 it had over 120 (Muckle, 2017).

Green alleys were not new: another borough (The Plateau) had begun transforming alleys into gardens in 1997, but they had been difficult to maintain. The key innovation in Rosemont-La-Petite-Patrie was to encourage and empower residents to take over their alleys (as opposed to the borough doing so). Borough Mayor François Croteau, defying some of Montreal's city by-laws, turned a blind eye to citizens rearranging back alleys *without* borough intervention. The sense of ownership this gave residents meant that the alleys were not only created, but have been maintained and developed over the years.[14]

This is an interesting example of one borough (Rosemont-La-Petite-Patrie) imitating an environmental innovation introduced in another, but altering the regulatory framework – by choosing not to apply regulations – in such a way that it gained momentum and sustainability. Rosemont's

---

14 Personal discussions with Mayor Croteau, 25 September 2019.

way of doing has now been imitated elsewhere, and in Rosemont itself green alleys have become so ubiquitous that social economy enterprises have emerged to help citizens manage them (Nature-Action Québec, 2021). The green alley initiative in Rosemont has led to a cascading series of innovations, with the borough introducing participatory budgeting in 2019 in view of helping citizens guide money towards the green spaces they desire.

Discussion and Conclusion: Can Municipal Innovation Make a Difference?

In this chapter I first describe the Schumpeterian innovation process, developed and applied in an economic context. Similarly to Swann (2014), I argue that it can also apply to non-market innovators such as municipalities and civil society organizations, but that goals, values, and assessment of innovation in non-market contexts need to be understood differently. The goal is not to generate or satisfy market demand. Rather, goals are negotiated, arrived at through discussion, consensus seeking, and/or vote. This approach requires that values be made explicit, weighed against each other, and that there be processes for reconciling incompatible values (Anderson, 1995).

Even in a public-service, social, or civil-society context, innovation goals should be set against available resources and costs. Those costs, also subject to negotiation and discussion, can be financial, cultural, environmental, equity-related, or time. Nevertheless, innovation in non-market contexts needs to be assessed before it is implemented (What are its ramifications and possible negative externalities?), as well as afterwards (Has it been successful? Were there unforeseen consequences?). A successful innovation will reach (or enable movement towards) goals without incurring excessive costs, however defined.

Innovation is shown to be political by considering non-market innovation, and by making it explicit that negotiations are necessary, that there are competing value systems, and that benefits and costs need to be weighed across complex and interrelated systems. The positive connotations that surround the term are stripped away when this complexity is faced. Not only are some innovations undesirable by almost any standards (yet may serve political purposes – innovative ways of committing genocide, for instance); many innovations are undesirable from some perspectives and according to some values, yet desirable from other perspectives and according to other values. For example, extracting oil from Alberta's tar sands was not possible before innovative extraction techniques were implemented. Yet whilst these innovations can be considered catastrophic from the perspective of the environment, they can be considered beneficial other perspectives, such as Canada's balance of trade. Much of Canada's current political debate is dominated by the fallout from these innovations: despite being commonly viewed as successful private-sector innovations (and hence as essentially value-free), these extractive technologies are in fact highly political. Furthermore, the regulatory framework that has allowed these innovations to develop is also political: whether these regulations were adapted (i.e., new regulations were necessary) or not (i.e., existing rules sufficed), a political choice was made and certain values were promoted above others.

This raises the issue of the nature of innovation in the public sector. Wagenaar and Wood (2018) suggest that it is political in essence, that public-sector innovation is almost wholly defined, introduced, and judged politically. Whilst their argument is compelling when applied to large-scale, ideologically motivated changes in governance and administration (such as, for example, innovations associated with New Public Management over the last 30 years), it is less clear that it also applies to smaller-scale innovations of the sort implemented by small municipalities and boroughs.

Within a municipal or borough, many innovations are problem-solving. Employees, administrators, politicians, or citizens identify a problem and/or something that can be improved; they then convince backers within the municipality that the envisaged solution deserves investigation, and a period of development and pilot testing follows. Of course, politics is involved, but if the problem is clearly defined, the objectives are discussed, and a method of evaluation is agreed upon, success or failure is more than rhetorical: it can be observed. Indeed, even New Public Management is now being assessed and found wanting on the basis of its own objectives (Hood & Dixon, 2015; Diefenbach, 2009; Casebourne, 2014; Bowman et al., 2015). Such assessments will usually be more immediate in local contexts, where the scope of innovation is smaller and where the population is in direct contact with decision-makers.

Another key aspect of innovations introduced by municipalities – and more generally by public bodies – is that they affect the lives of many citizens, whether they desire the innovation or not: the duty of care that public organizations have to the population means that innovation implementation should be more cautious than in the private sector. Furthermore, municipalities manage the immediate context (roads, parks, water, noise, local services, etc.) within which people live: innovation for its own sake must be set against continuity and stability. Continuity and stability – often downplayed in technophilic accounts of innovation (Borins & Herst, 2018) – are of concern to municipalities (and other public and community bodies) because unforeseen negative consequences cannot be dismissed as "externalities" (Swann, 2014). More often than not, they will remain internal to the municipality.

It is because municipalities are directly engaged with the immediate physical environment that their actions often affect environmental sustainability. Examples are given of some municipal and borough innovations that have enhanced sustainability.

An objection that can be made to the chapter's argument is that, however laudable the actions of these (and other) innovative municipalities, small municipal units cannot have much impact: only metropolitan, national, and global governance and regulation can tackle the major environmental challenges to the planet. This objection is similar to that levelled against Gibson-Graham's (1996) argument for local resistance to capitalism: they respond that perceiving capitalism as a monolith, only to be overturned by a commensurately monolithic revolution, can blind people to the small ways that local variants of capitalism can be altered, humanized, and eventually replaced. Their argument is not intended to disparage broader attempts to find alternatives to capitalism: rather, it seeks to empower smaller actors – individuals, civil organizations, and, why not, boroughs and municipalities. Global issues (of which climate change and environmental degradation are two) can be broken down and tackled – modestly, but with immediate and real effect – locally.

Municipalities and boroughs, whilst small, are not isolated. They are highly networked. Organizations such as the UMQ ensure that promising innovation in one Quebec municipality is shared with others. One criterion for entering the competition is willingness to share. Within large cities, boroughs imitate each other and share their innovations through networks of municipal actors. At a wider scale there is – and has been for many years – considerable policy transfer between boroughs and municipalities nationally and internationally. A good example is urban planning techniques and practices that have been shared among municipalities for over a century, as documented by Harris and Moore (2013).

Although Wills (2016) rightly points out that not all municipalities and boroughs have the same internal capacities, and although innovative local government practices will not be *sufficient* to bring about global change, they are a *necessary* component in the struggle for sustainability. In his book about governance in an era of increasingly complex socioecological systems, Young (2017) suggests that a successful system will call upon a wide variety of state, private, and other actors, and will build-in experimentation, nimbleness, and redundancy. Notwithstanding the importance of some global regulations and norms, a top-down centralized approach cannot tackle current socio-ecological problems. The type of municipal innovation I describe contributes to experimentation, nimbleness, and redundancy, increases the number of actors involved, and contributes to nudging the complex global system towards greater sustainability.

However, municipalities remain small actors at the national and global scale: even metropolitan areas have difficulty being heard on national and global stages. But perhaps "being heard" is not the way efforts of cities and municipalities will bear fruit. In a complex global socio-ecological system no single actor or level of government can have decisive impact (Young, 2017): yet each actor and level of government is important and can contribute to overall outcomes by increasing social capital, by reacting to local issues in a principled way, and by striving towards sustainable goals. In that sense, municipal initiatives are reactive and introduce new ways of doing. Furthering sustainable outcomes is critical to the overall project of stalling, and perhaps reversing, climate change, pollution, and environmental degradation.

REFERENCES

Andersen, E. (1994). *Evolutionary economics: Post-Schumpeterian contributions.* Routledge.

Andersen, E. (1995). *Values in ethics and economics.* Harvard University Press.

Bailly, G., & Coulbaut-Lazzarini, A. (2018). La réplication de démonstrateurs comme moteurs d'une nouvelle forme de production d'écosystèmes territoriaux? In D. Desponds & I. Nappi-Choulet (Eds.), *Territoires intelligents: un modèle si smart?* (pp. 101–122). Éditions de l'Aube.

Barras, R. (1986). Towards a theory of innovation in services. *Research Policy, 15*(5), 161–73. https://doi.org/10.1016/0048-7333(86)90012-0

Baudrillard, J. (1970). *La Société de Consommation.* Gallimard.

Bell, T. (2016). Digital innovation in the public sector: Why so slow? *Open Access Government.* https://www.openaccessgovernment.org/digital-innovation/24567/

Borins, S. (2014). *The persistence of innovation in government.* IBM Center for the Business of Government.

Borins, S., & Herst, B. (2018). *Negotiating business narratives.* Palgrave Pivot.

Bowman, A., Erturk, I., Folkman, P., Froud, J., Haslam, C., Johal, S., Leaver, A., Moran, M., Tsitsianis, N., & Williams, K. (2015). *What a waste: Outsourcing and how it goes wrong.* Manchester University Press.

Casebourne, J. (2014). *Why motivation matters in public sector innovation.* Nesta.

Casson, M. (2005). Entrepreneurship and the theory of the firm. *Journal of Economic Behavior & Organization, 58*(2), 327–348. https://doi.org/10.1016/j.jebo.2004.05.007

Chesbrough, H. (2003). *Open innovation: The new imperative for creating and profiting from technology*. Harvard Business Press.

Diefenbach, T. (2009). New Public Management in the public sector organizations: The dark sides of managerialistic "enlightenment." *Public Administration, 87*(4), 892–909. https://doi.org/10.1111/j.1467-9299.2009.01766.x

Dolowitz, D., & Marsh, D. (2000). Learning from abroad: The role of policy transfer in contemporary policy-making. *Governance, 13*(1), 5–23. https://doi.org/10.1111/0952-1895.00121

Drummondville. (2018). Service du développement durable et de l'environnement. http://www.drummondville.ca/service-municipal/service-du-developpement-durable-et-de-lenvironnement/

Elliott, S. (2014). Anticipating a Luddite revival. *Issues in Science and Technology, 30*(3), 27–36. https://issues.org/stuart/

Fioramonti, L. (2014). The world's most powerful number: An assessment of 80 years of GDP ideology. *Anthropology Today, 30*(2), 12–15. https://doi.org/10.1111/1467-8322.12098

Fourastié, J. (1979). *Les Trente Glorieuses*. Fayard.

Gibson-Graham, J. K. (1996). *The end of capitalism (as we knew it): A feminist critique of political economy*. Blackwell.

Godin, B. (2008). *Innovation: History of a category*, Project on the Intellectual History of Innovation, working paper 1, http://www.csiic.ca/PDF/IntellectualNo1.pdf

Grabher, G. (2018). Marginality as strategy: Leveraging peripherality for creativity. *Environment & Planning A, 50*(8), 1785–94. https://doi.org/10.1177%2F0308518X18784021

Greenfield, A. (2013). *Against the smart city*. Verso.

Harris, A., and Moore, S. (Eds.). (2013). Planning histories and practices of circulating urban knowledge. *International Journal of Urban and Regional Research, 37*(5), 1499–1583. https://doi.org/10.1111/1468-2427.12043

Hood, C. (1991). A public management for all seasons? *Public Administration, 69*, 3–19. https://doi.org/10.1111/j.1467-9299.1991.tb00779.x

Hood, C., & Dixon, R. (2015). *A government that worked better and cost less? Evaluating three decades of reform and change in UK central government*. Oxford University Press.

Howell-Moroney, M. (2008). The Tiebout hypothesis 50 years later: Lessons and lingering challenges for metropolitan governance in the 21st century. *Public Administration Review, 68*(1), 97–109. https://doi.org/10.1111/j.1540-6210.2007.00840.x

Kahn, M., Li, P., & Zhao, D. (2015). Water pollution progress at borders: The role of changes in China's political promotion incentives. *American Economic Journal: Economic Policy, 7*(4), 223–224. http://doi.org/10.1257/pol.20130367

Kastelle, T. (2015). How does innovation work in the public sector? In M. Stewart-Weeks & T. Kastelle (Eds.), Innovation in the public sector. *Australian*

Journal of Public Administration, 74(1), 63–72. https://doi.org/10.1111/1467-8500.12129

Ladouceur, E. (2006, 3 May). Deux autres distinctions pour Terrebonne. *La Revue.* https://numerique.banq.qc.ca/patrimoine/details/52327/2695911

Magnusson, W. (2005). Are municipalities creatures of the province? *Journal of Canadian Studies, 39*(2), 5–29, https://doi.org/10.3138/jcs.39.2.5

Mazzucato, M. (2015). *The entrepreneurial state.* Anthem Press.

Meijer, A., & Thaens, M. 2018. Urban technological innovation developing and testing a sociotechnical framework for studying smart city projects. *Urban Affairs Review, 54*(2), 363–87. https://doi.org/10.1177%2F1078087416670274

Montpetit, G. (2018, 8 February). Biométhanisation et bémols. *Le Courrier de Ste-Hyacinthe.* http://www.lecourrier.qc.ca/stha01-08-02-18-9013/biomethanisation-bemols

Montréal. (2016). *Sustainable Montréal 2016–2020: Together for a sustainable metropolis.* City of Montréal. https://numerique.banq.qc.ca/patrimoine/details/52327/2786004?docref=-Q1VTIBRskgdyxwB7O2PMA

Morozov, E. (2017, 22 October). Google's plan to revolutionise cities is a takeover in all but name. *The Guardian.* https://www.theguardian.com/technology/2017/oct/21/google-urban-cities-planning-data

Muckle, F. T. (2017). Plus de 100 ruelles vertes dans Rosemont-La-Petite-Patrie. 24 heures. https://www.24heures.ca/2017/08/27/plus-de-100-ruelles-vertes-dans-rosemont-la-petite-patrie

Nature-Action Québec. (2021). À propos. https://nature-action.qc.ca/a-propos/

O'Bryan, M. (2013, November). Innovation: The most important and overused word in America. *Wired.* https://www.wired.com/insights/2013/11/innovation-the-most-important-and-overused-word-in-america/

OECD. (2005). *Oslo manual.* Organisation for Economic Cooperation and Development.

Oliviera, M. (2018, 2 May). *Google's Sidewalk Labs project in Toronto raises privacy, data concerns.* Global News. https://globalnews.ca/news/4181324/toronto-sidewalk-labs-quayside/

Perry, J., & Hondeghem, A. (Eds.). (2008). *Motivation in public management: The call of public service.* Oxford University Press.

Potts, J. (2009). The innovation deficit in public services: The curious problem of too much efficiency and not enough waste and failure. *Innovation, 11*(1), 34–43. https://doi.org/10.5172/impp.453.11.1.34

Québec. (2018). *Information sur le réseau routier.* http://www.transports.gouv.qc.ca/fr/projets-infrastructures/info-reseau-routier/Pages/information-sur-le-reseau-routier.aspx

Rauch, D., & Schleicher, D. (2015). Like Uber, but for Local government law: The future of local regulation of the sharing Eeconomy. *Ohio State Law Journal, 76*(4), 901–964.

Sachs, J. (2018, 13 April). Forget Trans Mountain, here's the sustainable way forward for Canada's energy sector. *The Globe and Mail.* https://www.theglobeandmail.com/opinion/article-the-sustainable-way-forward-for-canadas-energy-sector/

SCESD. (2017). *Report to the House of Commons Standing Committee on Environment and Sustainable Development on the Federal Sustainable Development Act.* Ottawa: Government of Canada.

Scherer, F. (1986). *Innovation and growth: Schumpeterian perspectives.* The MIT Press.

Schumpeter, J. (1936). *The theory of economic development.* Harvard University Press. (Original work published 1911)

Schumpeter, J. (1942). *Capitalism, socialism and democracy.* Harper.

Schwartz, S. (2012). An overview of the Schwartz theory of basic values. *Online Readings in Psychology and Culture, 2*(1). https://doi.org/10.9707/2307-0919.1116

Serieys, J. (2018). *14 décembre 1789: Création des communes par la Révolution française.* Midi Insoumis, Populaire et Citoyen. www.gauchemip.org/spip.php?article10803

Seto, K., Golden, J., Alberni, M., & Turner, B. (2017). Sustainability in an urbanizing planet. *Proceedings of the National Academy of Sciences, 114*(34), 8935–8938. http://doi.org/10.1073/pnas.1606037114

Shane, S., & Cable, D. (2002). Network ties, reputation, and the financing of new ventures. *Management Science, 48*(3), 364–381. http://doi.org/10.1287/mnsc.48.3.364.7731

Shearmur, R. (2016). Debating urban technology: Technophiles, Luddites and citizens. *Urban Geography, 37*(6), 807–809. https://doi.org/10.1080/02723638.2016.1207914

Shearmur, R., & Poirier, J. (2017). Conceptualizing nonmarket municipal entrepreneurship: Everyday municipal innovation and the roles of metropolitan context, internal resources, and learning. *Urban Affairs Review, 53*(4), 718–751. http://doi.org/10.1177/1078087416636482

Solow, R. (1994). Perspectives on growth theory. *Journal of Economic Perspectives, 8*(1), 45–54. https://doi.org/10.1257/jep.8.1.45

Srnicek, N. (2016). *Platform capitalism.* Polity.

Swann, P. (2014). *Common innovation.* Edward Elgar.

Temenos, C., & McCann, E. (2012). The local politics of policy mobility: Learning, persuasion, and the production of a municipal sustainability fix. *Environment & Planning A, 44,* 1389–1406. https://doi.org/10.1068%2Fa44314

Tiebout, C. (1956). A pure theory of local expenditures. *Journal of Political Economy, 64*(5), 416–424. https://doi.org/10.1086/257839

Tissot, S. (2015). *Gentrifying diversity in Boston's South End.* Verso Books.

UN. (2018). *The sustainable development goals report: 2018.* United Nations.

Union des municipalités du Québec (UMQ). (2018). Trois-Rivières: Optimisation des processus liés à la gestion du plan triennal d'immobilisations et à la gestion contractuelle. https://umq.qc.ca/publication/trois-rivieres-optimisation-des

-processus-lies-a-la-gestion-du-plan-triennal-dimmobilisations-et-a-la-gestion-contractuelle/
Union des municipalités du Québec (UMQ). (2021a). Saint-Hyacinthe: Pionniére en biométhanisation. http://www.ville.st-hyacinthe.qc.ca/services-aux-citoyens/biomethanisation.php and https://umq.qc.ca/publication/saint-hyacinthe-pionniere-biomethanisation/
Union des municipalités du Québec (UMQ). (2021b). Trois-Rivières: Système informatisé de réparation de véhicule. https://umq.qc.ca/publication/trois-rivieres-systeme-informatise-de-repartition-de-vehicule/
Wagenaar, H., & Wood, M. (2018). The precarious politics of public innovation. *Politics and Governance, 6*(1), 150–160. http://doi.org/10.17645/pag.v6i1.1275
West, D., & Lu, J. (2009). *Comparing technology innovation in the private and public sectors*. Brookings Institute.
Wills, J. (2016). *Locating localism: Statecraft, citizenship and democracy*. Policy Press.
Young, O. (2017). *Governing complex systems: Social capital for the Anthropocene*. MIT Press.

PART TWO

The Role of Law and Overcoming Collective Action Problems

4 Sustainable Development and Property Rights: Citizen Participation in Dismantling Urban Environmental Regulation in British Columbia

DEBORAH CURRAN

Introduction: Urban Sustainable Development Facing Property Rights

Municipalities in British Columbia are leaders in Canada, with many taking an integrated sustainable community planning approach to land use bylaws and service delivery. Sustainable development has had a strong influence on local government approaches since 2009, when over 150 local governments signed onto the Climate Action Charter, committing them to reducing greenhouse gas (GHG) emissions, and the provincial government–mandated GHG reduction goals and activities in all community plans. From an environmental protection perspective, it is becoming the norm to consider parks and greenspace as integral elements of green infrastructure, and there is a move to include natural assets in capital accounting.

However, this chapter adds a cautionary note to the book themes of social movements, policy innovation, and the role of law. The opportunity and potential offered by planning and design as collective action responses to the wicked problems associated with sustainable development can fail if they neglect to shift dominant narratives and legal structures, such as private property rights, that uphold individual autonomy. Outside of sustainable development, even when urban and suburban local governments are behaving similarly by seeking growth, competing, and patenting their policy processes, as Shearmur notes in Chapter 3, those collective behaviours can be stymied by organized landowner resistance aimed at maintaining the illusion of individual autonomy.

Citizens and municipal governance processes in the District of Saanich, a suburban municipality adjacent to the City of Victoria, performed this tension between collective action through environmental bylaws and the perceived strength of individual autonomy expressed through the protection of private property rights. The District of Saanich has always been at the forefront of ecological protection. With one of the oldest urban growth boundaries in North America (Dawkins & Nelson, 2002), it built on this foundational element of sustainability by being the first to comprehensively map its ecologically sensitive areas and have those maps available to landowners when they sought approval for development (Curran & Krindle, 2016, pp. 95, 137–138). Buoyed by citizen support for protection of the natural environment, this transparency translated into development permit areas (DPAs) for protection of the natural environment, a unique overlay zone enabled by the *Local Government Act*, for riparian areas. Environmental development permit areas (EDPAs) require landowners to apply for a permit that may impose conditions relating to environmental protection or restoration when development occurs. In 2012 and supported by similar mapping, Saanich staff and council extended these EDPAs onto 2 per cent of non-riparian properties, which included properties with marine shorelines (Diamond Head Consulting, 2017). In a suburban context where development is predominantly low density, this policy innovation was one of the first times that a local government in British Columbia had protected non-riparian ecology using EDPAs.

However, as noted by Shearmur in Chapter 3, innovation always comes with a cost. Just three years later, citizens generated a 44-page rebuttal to the EDPA, launched a website, and paid for lawn signs declaring "Preserve property rights: Does the Saanich EDPA bylaw negatively impact your property?" Hundreds of citizens filled special council meetings demanding the repeal of the terrestrial EDPA as flawed in designation and process, an overextension of municipal regulation, and a "taking of property rights."

This story of the repeal of the Saanich EDPA is the flip side of citizen engagement in support of sustainable development when legally and historically ingrained cultural perceptions about "private property rights" and the autonomy that property ownership should afford comes up against long-term ecological health expressed through policy innovation. In an era of sustainable development, particularly in British Columbia, it highlights the need to keep site-specific regulation closely tied to long-term municipal policy, and to ensure that as political winds change, the commitment to foundational pieces of the regulatory apparatus are well understood. Analogizing to Manaugh and Dreszner's argument in Chapter 2 that the subject matter issues of social movements benefit from official political support, lack of support from current social movements can threaten policy infrastructure aimed at sustainability.

This chapter begins by laying out the factors contributing to the ascent of sustainable development in urban and suburban Canada. Tied closely to this "smart growth" is environmental protection and a permissive legal structure for the regulation of land where extensive limitations on private property are well accepted but where assertions of "private property rights" regularly occur. The chapter turns to the legal parameters of EDPAs as the case study, which can be characterized as extensive site-specific local government regulation that can require ecological restoration and dedication of watercourses. With this background, the chapter sets out the specific circumstances of the repeal of the Saanich EDPA in response to citizen critique. Finally, the chapter concludes with observations about the opportunity for sustainable development to recast what is often posed as a dichotomy between land regulation and private property as interdependence where both the state and property owners secure benefits for one another.

Sustainable Development and Property Rights in Canada

The release of the Brundtland Commission report (World Commission on Environment and Development, 1987), largely credited with popularizing the concept of sustainable development, coincided with the daylighting of the sustainable communities' movement in Canada. Writing from Vancouver, British Columbia, Roseland's 1992 *Building Sustainable Communities* was one of the first treatises on sustainable community development. Four factors, in particular, have influenced the uptake of urban sustainable development in Canada: the loss of agricultural land, the cost of sprawl, the move towards "smart growth," and the concern with GHG emissions and climate change.

First, as an early leader, and still the best practice in North America for the protection of agricultural land, the Province of British Columbia established the Agricultural Land Reserve (ALR) in the 1970s (Curran, 2019). A provincial zone for classes 1 to 5 soils under the Canada Lands Inventory, the ALR prohibits non-farm uses or subdivision of farmland unless a provincial commission gives permission (*Agricultural Land Commission Act*, 2002). The regime has decreased the conversion of scarce farmland – less than 4 per cent of the provincial landscape – to urban and non-farm uses from 9,000 hectares to 200 hectares per year, and the zone acts largely as an urban growth boundary in the fast-growing urban areas of Greater Vancouver and the Okanagan Valley (Curran, 2019).

Second, comprehensive studies on the cost of sprawl daylighted the long-term expense of business-as-usual development of single detached subdivisions and highway expansion. In particular, Blais and Slack's work on infrastructure costs (Blais, 2010; Slack, 2006) in Ontario influenced the provincial government's enactment of growth management laws to direct new development into serviced areas and protect agricultural and green space (Curran, 2019; *Greenbelt Act*, 2005; *Places to Grow Act*, 2005). At the same time, SmartGrowth BC operated in British Columbia for a decade as the only province-wide non-profit organization in Canada that focused on land use at a local scale. A pillar of the "smart growth" message was integrating urban development into ecosystem functions (Curran & Leung, 2000).

Third, while provincial and local governments explored the costs of sprawl, urban sustainable development began to manifest in North America as "smart growth," a comprehensive package of policy and infrastructure approaches that includes protection of the working landscape (agricultural and forest land), containing urban areas, fostering green infrastructure and preserving important ecosystems, creating compact, complete, inclusive communities that host a range of housing and transportation choices, linking land use and transportation decisions, and promoting community involvement in development decisions (Curran, 2003; Smart Growth Network, 2006). Environmental protection and preserving the working landscape of agricultural and forested lands are explicit components of smart growth and sustainable development. British Columbia preceded Ontario's Places to Grow regional approach to growth management with discretionary regional growth strategies, the purpose of which reads like a wish list for smart growth (*Local Government Act*, 2015, s. 428). An explicit example is the City of Vancouver linking density with environmental protection and smart growth in its EcoDensity Charter (City of Vancouver, 2008).

Finally, concern with GHG emissions and the anticipated impacts of climate change motivated all levels of government to mandate changes to community development. Most starkly, in 2007 the Province of British Columbia enacted a suite of laws designed to reduce GHG emissions. Two aspects of these laws pushed sustainable development at a local level. The Province amended the *Local Government Act* to require that all community plans contain targets and actions for the reduction of GHGs (*Local Government Act*, 2015 s. 473[3]). While the law did not set specific standards, it required local governments to turn their minds to GHG reduction and include proposed targets and actions in community plans. In addition, the *Greenhouse Gas Reduction Targets Act* (2007, now titled the *Climate Change Accountability Act*) mandated carbon-neutral government operations by 2010 (s. 5), which included public sector institutions like universities (s. 6). The Province invited municipalities to sign on to carbon neutrality through the voluntary Climate Action Charter, which committed them to being carbon neutral in corporate operations by 2012, measuring and reporting community GHG emissions, and creating compact, complete, energy efficient communities (Province of BC, Union of British Columbia Municipalities, and Signatory Local Governments, 2007). Most municipalities and regional districts in BC – 187 of 192 – have signed the Climate Action Charter (Province of BC, n.d.).

While these four factors influenced the uptake of sustainable development in Canada, it is important to note that it advanced with citizen support, particularly in British Columbia. In 2000, with the endorsement of community organizations across the province, several public sector and non-profit institutions launched a new non-governmental organization called Smart Growth British Columbia, the first province-wide land use organization in Canada. Modelled on the statewide "Friends of" and "Smart Growth" organizations in the United States, for a decade Smart Growth British Columbia provided a forum and policy for sustainable land use. It supported both community organizations and municipalities in sustainability planning.

At the same time, local governments and institutions began requiring a focus on sustainability in large-scale developments. The City of Victoria sold the 13-acre Dockside property in downtown on the harbour on the basis of an integrated sustainability plan. The developers committed to achieving the green building rating system Leadership in Energy and Environmental Design platinum standard for the whole neighbourhood or face up to a $1 million penalty (Curran, 2016). Likewise, when Simon Fraser University on Burnaby Mountain decided to build, it sought a rezoning to consolidate the allowable density at the top of the hill and required developers via a ground lease to meet innovative rainwater management, energy efficiency and water reduction standards (Curran, 2016). In contrast, the application of sustainable development to neighbourhoods of single detached housing lags behind other more systemic or large-scale efforts, and often generates considerable controversy. Residents typically resist densification in existing neighbourhoods, an example of which is the opposition to the City of Vancouver's EcoDensity initiative (CBC News, 2008; Rosol, 2013) or to laneway housing (Levy, n.d.).

The evolution of urban sustainable development in Canada has moved activities from the public sphere onto private land. Early attention was given to macro policy initiatives such as growth management, urban greenspace, public transportation, and greening municipal operations by reducing energy consumption and GHG emissions. Examples include the early adoption of green building rating systems for municipal buildings (City of Calgary, n.d.; City of Kingston, n.d.), and pilot projects on municipal property or using municipal funding. A recent pilot project involves the City of Prince George installing the first solar parking lot in British Columbia (Ballard, 2018). Some local governments, such as the City of Victoria with the Dockside Green neighbourhood, extended sustainable development to procurement or attached sustainability criteria to the sale of land (Curran, 2016). Others provide incentives for sustainable development through amenity density bonus (Curran & Kindle, 2016) or redevelopment tax incentives (City of Victoria, 2009). Finally, municipalities are requiring the inclusion of sustainable development infrastructure in high-density or large-scale commercial development. Examples include mandatory green roofs (City of Port Coquitlam, 2008; City of Toronto, 2017) and connection to a municipal- or neighbourhood-owned district energy systems (City of North Vancouver, n.d.; Curran, 2016).

Unlike in many countries (Alterman, 2010), this uptake in sustainable development occurs in a permissive regulatory context for land in Canada where property rights do not have constitutional protection and, therefore, "regulatory takings" are virtually non-existent. As courts have noted, "In this country, extensive and restrictive land use regulation is the norm. Such regulation has, almost without exception, been found not to constitute compensable expropriation" (*Mariner Real Estate Ltd. v. Nova Scotia (Attorney General)*, 1999, para 42).

This legal reality comes as a surprise to many landowners, as the Canadian approach to land use is to allow governments to restrict virtually all use of land by, for example, zoning regulation without compensating the property owner. This "no compensation" principle is codified in the municipal statutes of several provinces, which essentially state that no compensation will be paid for changes in the value of land caused by specified decisions made under a land use bylaw or permitting function (British Columbia *Local Government Act*, s. 458; Alberta *Municipal Government Act*, s. 621). It is only when regulation takes away virtually all incidents of private ownership that the regulation will be found to be improper. These "regulatory takings" or regulatory expropriations are few and far between in Canada. The successful "takings" cases in the United States do not have a corollary in this country; no court in Canada has ever found land use regulation to result in a regulatory expropriation. Courts have awarded compensation for "regulatory takings" related to loss of mineral rights upon the creation of a park (*R. v. Tener*, 1985; *Casamiro Resource Corp v. British Columbia (Attorney General)*, 1991) or for the removal of all economic viability, including goodwill (*Manitoba Fisheries Ltd v. The Queen*, 1979), but not for limitations on the use of land.

In the land use context, Canadian courts note that local governments may change zoning up or down to realize legitimate public interests without attracting liability to compensate landowners for changes in property values (*Toronto Area Transit Operating Authority* v. *Dell Holdings*, 1992). The duty to compensate does not arise when municipalities deny subdivision or rezoning applications, or use other growth management mechanisms. This holds true even where development is limited in anticipation of future acquisition for public purposes (*Vancouver v. Simpson*, 1977; *Calgary (City) v. Hartel Holdings*, 1984). Courts are clear that devaluation of land does not equal the loss of an interest in land for which compensation should be paid (*Manitoba Fisheries*, 1979; *Tener*, 1985). Likewise, landowners affected by development freezes do not have a right to compensation, in part because such moratoria do not take an incident of property and confer it on the government, which is an element of the legal test for regulatory takings (*Calgary v. Nilsson*, 1999).

Courts draw a clear distinction between the exercise of land use jurisdiction and other governmental regulation to the extent that landowners "whose land is caught up in a zoning or planning process but not expropriated must simply accept in the public interest any loss that accrues from delay" (*Toronto Area Transit*, 1992, para. 52).

This legal landscape of significant local government authority to regulate the use of land without needing to compensate landowners for loss of value or use has its critics. More broadly, the separation between state and private responsibilities (Epstein, 1985) and the "property rights movement," in part organized against environmental regulation, is well documented in the United States (Marzulla & Marzulla, 1997). Closer to home, the critique is more muted and focuses on the arbitrariness of local government land use decisions such as zoning (Morgan, 2012), and, more generally, the loss of economic potential of private property due to government regulation (Milke, 2012).

The persistence of the private property narrative is remarkable, given the longstanding focus on sustainable development in Canada and the broad legal authority for restricting the use of private property. These two divergent approaches are illuminated by property-specific environmental protection mechanisms called EDPAs. An extension of more onerous provincial regulation in riparian zones, EDPAs permit local governments to impose environmental protection conditions on new development that can fundamentally shape where owners build and what ecological features they must protect.

Environmental Development Permit Areas: Law and Practice

Development permit areas are a property-specific land use regulation that local governments may designate for a variety of purposes, such as to manage hazardous conditions, environmental protection, the form and character of commercial development, and energy and water efficiency. A DPA is akin to an overlay zone (Blackwell, 1989). It creates an additional requirement for a permit that imposes site-specific conditions on new development and redevelopment.

The authority for a local government to designate a DPA for protection of the natural environment is exclusive to the province of British Columbia in Canada. A search of municipal planning statutes revealed no other provincial or territorial jurisdiction that enables local governments to provide this type of site-specific attention to the ecological impact of development. Although all provinces enable municipal protection of the natural environment through community planning, no other province or territory authorizes fine-grained development permitting that can address protection of the natural environment on a property-by-property basis.

Pursuant to the British Columbia *Local Government Act*, local governments may designate EDPAs in a community plan or zoning bylaw for the "protection of the natural environment, its ecosystems, and biological diversity" (s. 488[1][a]). The community plan must specify guidelines by which staff may address the objective or special conditions that justify the EDPA (s. 488[2]). An EDPA designation prohibits subdivision, construction or alteration of a building or other structure, or land alteration unless an owner has obtained a development permit (s. 489). The conditions of a development permit may do any of the following (s. 491[1]):

(a) specify areas of land that must remain free of development, except in accordance with any conditions contained in the permit;
(b) require specified natural features or areas to be preserved, protected, restored or enhanced in accordance with the permit;
(c) require natural water courses to be dedicated;
(d) require works to be constructed to preserve, protect, restore or enhance natural water courses or other specified natural features of the environment;
(e) require protection measures, including that vegetation or trees be planted or retained in order to
   (i) preserve, protect, restore or enhance fish habitat or riparian areas,
   (ii) control drainage, or
   (iii) control erosion or protect banks.

When one considers that EDPAs apply to private land, it is clear that the provincial legislature provided local governments in British Columbia with comprehensive legal authority to protect the environment. This power, at the same time, can have a significant impact on private property. Two aspects of this jurisdiction are particularly noteworthy. The first is that the designation of EDPAs is for protection of the natural environment, its ecosystems, and biological diversity. On a broad and purposive reading of the law (*Nanaimo (City) v. Rascal Trucking*, 2000; *United Taxi Drivers Fellowship v. Alberta*, 2004), local governments can designate EDPAs across virtually their entire landscape as all parts of a landscape, including urban, can contribute to protecting ecosystems and biological diversity. Second, the possible permit conditions that staff may impose can extensively limit development activities on private land, require proactive restoration, and even transfer riparian property to a municipality or third party for protection. A property that is degraded may be required to be restored to a better post-development ecological condition.

Taken to their most creative limits, EDPA permits can transform the way typical development occurs. This approach is not without precedent in British Columbia, where the provincial *Riparian Areas Regulation* under the *Riparian Areas Protection Act* requires all development within 30 metres of a fish-bearing watercourse to undergo similar ecological assessment and permitting. However, EDPAs are the most comprehensive environmental protection tool in non-riparian areas.

Since their enactment, no court has overturned an EDPA regime (Britton-Foster et al., 2016). Courts uphold DPA schemes where the local government elected officials have designated the DPA on reasonable evidence and with reasonably discernible boundaries (*Denman Island Local Trust Committee v. Ellis*, 2005). The guidelines act as "directing principle" (*48 Fr Hwy Land Ltd v. Langley (Township)*, 1999; *Rocky Point Metalcraft Ltd v. Cowichan Valley (Regional District)*, 2012) for how the local government will address the special conditions of the DPA but must provide some flexibility (*Washi Beam Holdings Corp v. West Vancouver (District)*, 1999; *511784 BC Ltd. v. Salmon Arm (District)*, 2001). Case law has typically upheld municipal staff and council decisions made pursuant to DPA regulations, but council members must show that they evaluated the application for a development permit pursuant to the guidelines in the community plan (*0742848 BC Ltd. v. Squamish (District)*, 2011). The few cases that have overturned a permitting decision under a DPA regime have found that council acted *ultra vires* by exceeding its jurisdiction by giving weight to factors not listed in its DPA guidelines (*Loewen v. Coquitlam (City)*, 1999; *Westfair Foods Ltd. v. Saanich (District)*, 1997). Finally, when landowners have challenged EDPA permits as "sterilizing" a portion of the land or amounting to a regulatory expropriation, courts have rejected those characterizations, finding that permits that locate dwellings in particular areas on a property do not change the density or zoning of the parcel (*0742848 BC Ltd.*, 2011). Reasonable permit conditions that are congruent with the purpose of the EDPA are not regulatory expropriation (*Bignell Enterprises Ltd. v. Campbell River (District)*, 1996).

In summary, in an era of sustainable development at the local level, extensive land use regulation of private property is well accepted in Canada and expressed through general no-compensation rules as well as broad authority for site-specific development controls. Courts are highly deferential to local governments' implementation of EDPA regimes and the conditions imposed in permits. Due to provincial government mandates for riparian areas, most EDPA regimes focus on watercourses (Curran & Krindle, 2016), but a few now apply into upland areas. It is this extension into upland ecosystems that aroused citizen and stakeholder action for the protection of their "property rights" that resulted in the repeal of the EDPA in the District of Saanich, just north of Victoria, British Columbia.

Case Study: The Rise and Fall of Environmental Development Permit Areas in Saanich, BC

The District of Saanich has always been a leader in growth management and urban environmental protection. It established the oldest urban growth boundary in North America in 1968 by creating water and sewer servicing limits along Wilkinson Road that, over time, solidified into a growth management and sustainable communities policy (District of Saanich, 2008, 4-10-4-11). The District created a Sensitive Ecosystems Atlas in 1996 to map the presence of sensitive ecosystems throughout the District on both public and private land. It was used as a resource for staff and landowners in the context of applications for development. In 2004 the Provincial Government enacted the Riparian Areas Regulation (RAR) that required local governments to verify that an ecological assessment had taken place for any land development proposals adjacent to fish-bearing streams. The RAR requires the assurance of a registered professional that the development will not result in harm to fish or fish habitat, or else senior government approval is required. The District implemented the RAR by establishing a riparian DPA and took an approach that was unique and helpful to landowners of pre-mapping the riparian area such that landowners could rely on that public assessment to site development or could hire their own registered professional.

In 2012, after further ground-truthing of the Sensitive Ecosystems Atlas, the District of Saanich created a non-riparian EDPA for 3,618 properties, some 2 per cent of parcels in the municipality (District of Saanich, 2016a). Throughout that year, staff created many public input

Sustainable Development and Property Rights 105

Figure 4.1 Picture of signs erected on private lawns as well as public boulevards and street crossings warning landowners about the EDPA. Photo by Deborah Curran.

opportunities, which included open houses, newspaper announcements, presentations, an online feedback form, and a project webpage (District of Saanich, 2016a).

In 2015, an organization called Saanich Citizens for a Responsible EDPA began publicly criticizing the EDPA regime and its application. In addition to a 44-page report (2015c), the group created websites (2015a), information sheets (2015b), lawn signs (see Figure 4.1), and media attention (Smart, 2015a, 2015b) to generate interest in the impact of the EDPA. This public information focused on limitation of property rights, decreasing land values, and the inadequate scientific or technical basis of the EDPA. Some of the information perpetuated inaccuracies about the EDPA and its impact. For example, the opening paragraph from a fact sheet from spring 2015 stated,

> The District of Saanich has imposed an Environmental Development Permit Area (EDPA) bylaw on over 2000 properties in Saanich, and this number is growing. *Your property may be in the EDPA or will soon be.* This bylaw restricts property owners from disturbing the soil or vegetation on their property. This can mean, among other things, no gardening, no fencing, no pathways, no pruning or brush cutting, no sheds or playhouses, and no farming (non ALR). In many instances, part or all of existing homes have been placed under the EDPA, further limiting the property value and use. *The restrictions are significant and most property owners are unaware.* (Emphasis in original)

District of Saanich staff had responded to concerns about the EDPA with a "Frequently Asked Questions" document that confirmed there would be no expansion of the EDPA and clarified how residential activities, such as gardening, are affected by the EDPA. In response, the Citizens for a Responsible EDPA dissected the District's responses, characterizing many of the District's responses as inadequate and misleading, taking issue with the mapping used to designate the EDPA and the characterization of terminology used by the District (Saanich Citizens for a Responsible EDPA, 2015b). Other statements by residents in the media, such as that they are not permitted to mow their lawns (Hutchinson, 2016), underscored the lack of understanding of EDPA regulation.

In response, staff submitted a report to the Natural Areas Advisory Committee to the District, a citizen committee, suggesting clarifying amendments to the EDPA. The Committee further recommended, and council moved forward with, a public process, given the increasing citizen concern about the EDPA. This public check-in proceeded from 29 May 2015 to 26 February 2016. While the District found that the response to the check-in did not garner more attention from the public than similar planning or development processes (District of Saanich, 2016a), for a policy that affects 2 per cent of properties the numbers seem remarkable. Overall, the District received 861 instances of written or verbal feedback from 498 distinct individuals or organizations and engaged in individual consultations with over 400 residents. The 225 formal feedback forms completed by property owners of land in the EDPA represented only 6 per cent of EDPA owners and 9 per cent of EDPA properties. The two open houses attracted over 560 attendees, with the town hall meetings hearing from 100 speakers.

The summaries and minutes from the town hall meetings and special council meeting on the EDPA reveal omnipresent attention to the impact of the EDPA as environmental regulation on property values, and the unfairness to property owners (District of Saanich 2015, 2016b, 2016c). Although a significant number of citizens spoke in favour of the EDPA, presenters focused on property rights and values, scientific inaccuracies of the EDPA, and fairness, consistent with the Saanich Citizens for a Responsible EDPA report. Related to the submissions about diminution in property rights and values, citizens expressed their view that the cost burden for environmental protection should be shared by all, and that if Saanich wants to protect the environment the municipality should pay for it.

Citizens also emphasized the alleged scientific inaccuracies of the EDPA designation, citing erroneous, outdated, and unprofessional mapping. Some asserted that staff created the regulation absent of scientific evidence or sound ecological criteria, noting that "applying a designation of highest biodiversity to houses, lawns, gardens and farm fields seems like a scientific oxymoron" (District of Saanich 2016b, p. 1). Many citizens expressed their opinion that properties should not be designated on maps using aerial photography, but that the conditions on any property should be ground-truthed. There was considerable support for taking decision-making away from staff and creating an unbiased, independent, non-academic scientific committee to report directly to the Mayor and council.

Fairness arguments focused on the application of the EDPA to only some properties and not municipal property, and staff discretion to interpret the guidelines as applied to individual properties. The sentiment was that property owners with land designated as EDPA had an unfair burden placed upon them for environmental protection that benefitted the entire citizenry. This unfairness was heightened because Saanich parks and other public lands were not subject to EDPA rules, and, in some areas, invasive species flourished on public land. Finally, the EDPA guidelines granted staff too much discretion to apply the guidelines in multiple ways to different properties. Citizens called into question the fairness of staff's application of the regulation, and the lack of specificity in the guidelines that would direct staff's actions more closely.

Finally, citizen input paid considerable attention to a lack of communication between Saanich and affected property owners. Overall, citizens said that the EDPA regime was poorly explained by staff, and that they were given little notice of its potential impacts at the time council enacted it.

A consultant hired by the District of Saanich to review the EDPA characterized the public process as "an acrimonious social discourse":

> It has become evident to our team through engagement that the District, public, developers and environmental professionals will need to make efforts to rebuild trust in the EDPA Bylaw and process to move forward with the protection of ESAs [environmentally sensitive areas] on private property, and that this need will not be resolved solely by implementing the recommendations of this review. A number of people highlighted their reluctance to engage on the topic of the EDPA, given the current type of acrimonious social discourse taking place, and the detrimental effect it is having amongst community members. This will be limiting to the ability for the District to engage with its community members. (Diamond Head, 2017)

Other commentary noted the frenetic nature of the debate:

> [L]ately, Saanich has become synonymous with squabbling. Bitter arguments. Finger pointing. Endless town-hall sessions. From at least one of the district's most prized neighbourhoods, there have been cries for secession, calls to join an adjacent municipality. A special district council meeting was held Wednesday night; it lasted well past midnight and solved nothing. According to residents, the problem is property rights. Specifically, their erosion. (Hutchinson, 2016)

While the number one response from all respondents was acknowledgment of the importance of protecting natural areas and the need to restore and link ecosystems in the future, other most frequent responses related to the devaluation of property, the inaccuracy of the mapping, and the perceived unfairness of the process the District used to enact the EDPA, lack of understanding of the EDPA, and impacts on property rights (District of Saanich, 2016a). Of note, the 13th most frequent response was opinions that include "coarse language, references to fascism or clandestine activities" (District of Saanich, 2016a, p. 8).

The municipal EDPA process continued over the next two years and included a complaint to the Office of the Ombudsperson of British Columbia (Dedyna, 2016), and a report from the British Columbia Assessment Authority finding no evidence that the EDPA devalued properties (2016). However, the death of a pro-EDPA Councillor (DeRosa, 2017a) and election of a new Councillor who stated clearly during her election campaign that the EDPA should be dismantled (DeRosa, 2017b) secured the votes for EDPA repeal. When council repealed the EDPA on 23 April 2018, they also passed a motion to develop a biodiversity conservation strategy to establish the overarching vision for ecological health in the municipality. Even though sensitive ecosystems and land adjacent to parks on private lands is currently unprotected, the municipality is revisiting the larger policy and decision-making framework guiding municipal environmental protection and bylaw activities. It is difficult to envision, given the degree of hostility from citizens towards municipal staff and council, how a broader ecological health scaffold can reintroduce the expression of public goods, such as habitat restoration, on private land.

Citizen Participation in Privatizing Environmental Protection or Mandating Public Sustainable Development

While sustainable development may be omnipresent and well incorporated into urban local governance, when confronted by the metanarrative of private property rights, it is vulnerable to dismantling. Even though there are trends in higher land values due to sustainable developments (Krause & Bitter, 2012), the perceived limitation of private property rights by local governments for environmental protection overrules personal interest in that value premium. This unconscious choice was repeated time and again in the Saanich EDPA example as landowners critiqued the regime on largely technical and economic grounds. While there are many observations – or, from the perspective of local governments, lessons – one can make about this case study, what stands out most strongly is how the citizens relied primarily on a technical, non-legal attack on the EDPA approach to support the overall call to protect "private property rights." As Saanich was acting within its jurisdiction, and thus the law, residents discredited the EDPA regime by pointing out how the mapping created absurd results, such as requiring the protection of lawn or a garage.

Residents also spent considerable effort demeaning staff expertise and their exercise of professional judgment about what ecosystem features must be protected and restored under the guise of "fairness." The tenor of the comments was that the regulations themselves were unfair because they applied only to some properties, and that staff discretion to apply the regulations to the properties within the EDPA was also unfair. These fairness arguments fail to grasp a central feature of municipal administration and decision-making – that much of municipal law involves permissible discrimination between different circumstances, properties, businesses, and areas to achieve the public purpose that municipal council is addressing. The heart of this fairness complaint is, in effect, the core of municipal discretion.

While access to information is a hallmark of deliberative democracy, when it comes down to impacts on specific parcels of private property, less may be more. Saanich's choice to provide high-quality mapping and save citizen taxpayers' money on ecological assessments ended up creating conflict. Landowners discredited the mapping by showing where it was not precisely accurate, and that was the basis on which council agreed to repeal the EDPA. Of note is that the precision demanded by landowners is not a legal requirement and does not reflect the discretion accorded to municipal decision-making.

When bylaws and policies apply in a site-specific way to private land, it may be that less information about the precise application to any parcel may be preferred, both as an invitation to a conversation and to neutralize the empathy associated with the specific impacts of policy regimes as collective action. The absence of site-specific detail requires landowners or their agents to have a conversation with local government staff to understand the intent, purpose, and parameters associated with the regulation. Other EDPA regimes in BC rely more heavily on landowners to assess the ecological values on their properties, rather than supplying pre-existing municipal mapping (Britton-Foster et al., 2016). This requires some back-and-forth between the registered professional carrying out the assessment, the landowner, and local government staff, and can enhance citizen comprehension of the application of broader sustainable development policy.

Baseline, rather than detailed, information can also maintain the overarching focus of the policy regime – in this case environmental protection – rather than the regime itself engaging at the level of site-specific impacts. While municipal bylaws are inherently discriminatory, as they make distinctions between categories of land and people by design, the individual and "discriminatory" application of any policy process can attract fairness critiques that lead to discrediting the entire policy platform. Generalized regulations that capture more properties and people in the restrictions aimed at achieving sustainable development goals appear more fair because they share the burden and benefit of the collective action through policy.

Similarly, applying environmental protection on private land as part of sustainable development is better served through the wide application of general policies that provide exemptions, rather than narrow application on only those properties where identified sensitive ecosystems are proven. Many local governments take this approach with their EDPA regimes (Britton-Foster et al., 2016), which allows the application of environmental protection on private land to most or all properties, rather than singling out just some properties at the outset. The impact is felt on most properties, except for those on which no ecological elements remain, linking all landowners into the ecological health policy commitment.

Wide application also makes restoration a viable conversation. When relying on identified sensitive ecosystems, property owners spend considerable time proving that the mapping is incorrect because no sensitive ecosystem remains. Restoration, as part of adaptation to climate change, is not well accepted as the role of private property. While ecological protection has some traction, restoration of degraded landscapes is seen as a task of public property. Restoration has some applicability to all properties, not just those containing sensitive ecosystems or rare plants or animals, thus sharing the responsibility for ecosystem health amongst all landowners.

More systemically, absent broader well-entrenched and accepted sustainability policy, such as urban growth boundaries that are expressed through community plans, ecological health strategies, zoning bylaws, and subdivision servicing policies, environmental protection policies on private land will be subject to intense and technical criticism that aims to pick holes in the program. It could be argued that citizens viewed the EDPA as a hegemonic strategy by the municipality, much like many residents in Vancouver saw the EcoDensity initiative (Rosol, 2013). Unless nested within a broader environmental program that has widespread citizen support, site-specific ecological protection will attract "private property rights" criticisms. An example of such a policy is Saanich's urban containment boundary that restricts water and sewer service to non-urban areas. No densification can occur in these areas, and that position is well accepted, although it affects the property value of rural landowners directly. Landowners need to see that the restriction on their use of property or enjoyment of land is applied widely, and that the local government itself abides by the same or more stringent rules. A biodiversity conservation or ecological health strategy can make the link on how ecological connectivity and site-specific restoration through EDPAs fit in with the entire context of watershed function.

Ultimately, however, macro- to site-specific policies and bylaws are insufficient, as the Saanich case study demonstrated. It is the sustainable development culture of a community and local government – a culture of conservation – that supports enduring ecological protection. Municipalities may have a regulatory function for environmental protection but, absent additional resources for continued interaction with the public and citizen owners, policy implementation will fail. Saanich had excellent mapping and a regulatory structure but did not dedicate resources for implementing private land stewardship programming and best practices pilot projects, nor did it interact with the landowning community to continue to explain the environmental protection program in a meaningful way.

Such an approach expands the concept of "governing through owners" or "mutual dependence," where "the state depends on owners to perform core functions just as owners depend on the state to protect their rights" (Katz, 2012, p. 2053). Land is no longer simply a square of which the owner has quiet enjoyment but a component of the health of a larger regional ecosystem. Owners would have responsibilities to maintain land as part of a "life support system" (Norton, 2011; Freyfogle, 2003). From a landowner perspective, strengthening or building customs around the permeability of property boundaries for conservation can shift practices to meet emerging ecological needs (Yung & Belsky, 2007). On the local government side, a place to start would be with regional and municipal plans that establish the policy direction for healthy ecological function such that when the bylaws that implement that overall policy limit the use of land, citizens can understand their role in the larger project of ecosystem function. This culture of conservation locates protection and restoration of the ecological conditions on private lands as part of the larger sustainability effort, of which citizen participation is a core.

Conclusion

In an era of strong urban sustainable development in British Columbia over the past 20 years, considerable progress has been made in the larger cities of Vancouver and Victoria on green energy, transportation, and increasing density in already urbanized areas, particularly redevelopment of downtown cores. However, that progress is challenged in a suburban context by landowners of single detached housing who seek to keep their neighbourhoods as low-density residential and who resist site-specific land use regulation enacted in the public interest that restricts their use of land in support of ecological values. Citizens rely on the legally weak notion, in Canada, of "private property rights" to resist participating in the regional project of watershed health. They also discredit narrowly applied regulatory and policy regimes enacted to achieve sustainable development goals by focusing on the unfairness of individual impacts of that collective action policy when provided with detailed information on the scope of that regime. This discrediting is in contrast to more broad but generally applied regulations where the sustainable development "costs" are distributed across a wider landscape.

It is unlikely that any local government, aided by land trusts, will ever have sufficient funds to secure as public land adequate ecological connectivity across the landscape, particularly given historic patterns of land development, subdivision, and transportation infrastructure. Sustainable development requires that environmental protection move beyond the public sphere of parks and onto private lands to bolster the urban forest, biodiversity, and riparian connectivity. One can see this movement in the focus on green roofs, which local governments favour because of their rainwater retention and management benefits (Sims et al., 2016). They can also provide significant support for biodiversity in an urbanized landscape (Oberndorfer et al., 2007).

The demise of the Saanich EDPA provides many lessons in local government and citizen behaviour. To move beyond the dichotomy of public regulation versus private property, a culture of conservation would require equal attention be paid to supporting the conception of private property as one aspect of cooperative community goods – or collective action – on multi-scalar – regional to local – plans and bylaws that set the framework for sustainable development.

Acknowledgment

Special thanks to Andie Britton-Foster and Gabrielle Grant, who comprehensively researched the law of environmental development permits in the spring of 2016 as part of their student work with the Environmental Law Centre at the University of Victoria.

REFERENCES

0742848 BC Ltd. v. Squamish (District) (2011) BCSC 747, 84 MPLR (4th).

48 Fr Hwy Land Ltd. v. Langley (Township) [1999] BCJ No 1861, 4 MPLR (3d) 53 (SC).

511784 BC Ltd. v. Salmon Arm (District) (2001) BCSC 245, 19 MPLR (3d) 232.

Agricultural Land Commission Act S.B.C. 2002 c. 36.

Alberta Municipal Government Act R.S.A. 2000 c. M-26.

Alterman, R. (2010). *Taking international: A comparative perspective on land use regulations and compensatory rights.* American Bar Association.

Ballard, J. (2018, 2 July). Prince George installs BC's first solar power parking lot. CBC News. http://www.cbc.ca/news/canada/british-columbia/prince-george-installs-b-c-s-1st-solar-power-parking-lot-1.4731263

Bignell Enterprises Ltd. v. Campbell River (District) [1996] BCJ No 1735, 34 MPLR (2d) 193 (SC).

Blackwell, R. J. (1989). Overlay zoning, performance standards, and environmental protection after Nollan. *Boston College Environmental Affairs Law Review, 16*(3), 615–660.

Blais, P. (2010). *Perverse cities: Hidden subsidies, wonky policy and urban sprawl.* UBC Press.

British Columbia Assessment Authority. (2016, 14 January). *District of Saanich Environmental Development Permit Area* (Victoria: BC Assessment.

British Columbia Local Government Act R.S.B.C. 2015 c. 1.

Britton-Foster, A., Grant, G., & Curran, D. (2016). *Environmental development permit areas: In practice and in caselaw.* Environmental Law Centre.

Calgary (City) v. Hartel Holdings, [1984] 1 S.C.R. 337, 53 N.R. 149, 31 Alta. L.R. (2d) 97, [1984] 4 W.W.R. 193, 53 A.R. 175, 8 D.L.R. (4th) 321, 25 MPLR 245, 8 Admin. L.R. 231 (SCC).

Calgary v. Nilsson, [1999] A.J. No 645.

Casamiro Resource Corp v. British Columbia (Attorney General), 1991 CanLII 211 (BCCA).

CBC News. (2008, 10 June). Controversial EcoDensity Charter passes in Vancouver. http://www.cbc.ca/news/canada/british-columbia/controversial-ecodensity-charter-passes-in-vancouver-1.745555

City of Calgary. (n.d.). Sustainable building policy. https://www.calgary.ca/cs/iis/green-building/calgarys-sustainable-building-policy.html

City of Kingston. (n.d.). LEED: Green building information. https://www.cityofkingston.ca/residents/environment-sustainability/climate-change-energy/green-buildings/leed-green-building-information

City of North Vancouver. (n.d.). Lonsdale Energy Corporation. https://www.cnv.org/city-services/lonsdale-energy

City of Port Coquitlam. (2008, 14 April). A zoning bylaw for the City of Port Coquitlam. Bylaw No. 3630.

City of Toronto. (2017, 9 November). Toronto Municipal Code Chapter 492 Green Roofs.

City of Vancouver. (2008). EcoDensity Charter. City of Vancouver.

City of Victoria. (2009, 25 June). Revitalization Tax Exemption (Green Power Facilities) Bylaw. Bylaw no. 09-040.

City of Victoria. (2016). Green development: New entanglements of property, planning and the Public Interest. In A. Smit & M. Valiante (Eds.), *Public interest, private property: Law and planning policy in Canada* (pp. 166–191). UBC Press.

Curran, D. (2003). *Smart growth summary.* West Coast Environmental Law Association.

Curran, D., & Krindle, E. (2016). *Green bylaws toolkit for conserving sensitive ecosystems and green Infrastructure* (2nd ed.). Wetlands Stewardship Partnership.

Curran, D., & Leung, M. (2000). *Smart growth: A primer.* Smart Growth BC.

Dawkins, C.J., & Nelson, A.C. (2002). Urban containment policies and housing prices: An international comparison with implications for future research. *Land Use Policy, 19*(1), 1–12. https://doi.org/10.1016/S0264-8377(01)00038-2

Dedyna, K. (2016, 29 March). Complaint over Saanich eco-bylaw thrown out. *Times Colonist.* http://www.timescolonist.com/news/local/complaint-over-saanich-eco-bylaw-thrown-out-1.2218611

Denman Island Local Trust Committee v. Ellis, (2005) BCSC 1238.

DeRosa, K. (2017a, 19 March). Obituary: Saanich Coun. Vic Derman, 72, hailed for work on the environment. *Times Colonist.* http://www.timescolonist.com/news/local/obituary-saanich-coun-vic-derman-72-hailed-for-his-work-on-environment-1.12211353

DeRosa, K. (2017b, 15 October). How Karen Harper could change Saanich Council. *Times Colonist.* http://www.timescolonist.com/news/local/how-karen-harper-could-change-saanich-council-1.23064944

Diamond Head Consulting. (2017). *District of Saanich Environmental Development Permit Area independent review.* District of Saanich.

District of Saanich. (2008, 8 July). Sustainable Saanich: Official community plan. Appendix A to Bylaw 8940.

District of Saanich. (2015). Summary of the Town Hall Meeting held in the Garth Homer Centre 813 Darwin Avenue Thursday November 12, 2015 at 7:00 pm.

District of Saanich. (2016a). *EDPA public process report.* District of Saanich.

District of Saanich. (2016b).Minutes of the Special Council Meeting held in the Fieldhouse G.R. Pearkes Recreation Centre, 3100 Tillicum Road Wednesday March 16, 2016 at 7:00 pm.

District of Saanich. (2016c). Summary of the Town Hall Meeting held at the Field House at Pearkes Recreation Centre 3100 Tillicum Road Thursday February 11, 2016 at 6:00 pm.

Epstein, R.A. (1985). *Takings: Private property and the power of eminent domain.* Harvard University Press.

Freyfogle, E.T. (2003). *The land we share: Private property and the common good.* Island Press/Shearwater Books.

Greenbelt Act. (2005). SO 2005, c. 1.

Greenhouse Gas Reduction Targets Act. SBC 2007 c. 42.

Hutchinson, B. (2016, 17 March). Livid BC homeowners say bylaw to protect "sensitive ecosystems" destroying their property values. *National Post.* https://nationalpost.com/news/canada/livid-b-c-homeowners-say-bylaw-to-protect-sensitive-ecosystems-destroying-their-property-values

Katz, L. (2012). Governing through owners: How and why formal private property rights enhance state power. *University of Pennsylvania Law Review, 160*(7), 2029–2059.

Krause, A.L., & Bitter, C. (2012). Spatial econometrics, land values and sustainability: Trends in real estate valuation research. *Cities, 29*, S19–S25. https://doi.org/10.1016/j.cities.2012.06.006

Levy, E.P. (n.d.). Vancouver laneway housing – the good, the bad and the ugly. https://lanewayhousing.wordpress.com.

Loewen v. Coquitlam (City), [1999] BCJ No 2167, 5 MPLR (3d) 135 (SC).

Manitoba Fisheries Ltd. v. The Queen, [1979] 1 SCR 101.

Mariner Real Estate Ltd. v. Nova Scotia (Attorney General), (1999) 177 D.L.R. (4th) 696, 68 L.C.R. 1, 90 A.C.W.S. (3d) 589, 178 N.S.R. (2d) 294, (N.S.C.A).

Marzulla, N.G., & Marzulla, R.J. (1997). *Property rights: Understanding government takings and environmental regulation.* Government Institutes.

Milke, M. (2012). *Stealth confiscation: How governments regulate, freeze, and devalue private property – without compensation.* The Fraser Institute.

Morgan, E. (2012). The sword in the zone: The fantasies of land-use law. *University of Toronto Law Journal, 62*(2), 163–199. https://doi.org/10.3138/utlj.62.2.163

Nanaimo (City) v. Rascal Trucking Ltd., [2000] 1 S.C.R. 342.

Norton, R.K. (2011). Reconciling sustainability with private property rights in planning law and policy: A review of Takings International, by Rachelle Alterman. *McGill International Journal of Sustainable Development Law and Policy*, 89.

Oberndorfer, E., Lundholm, J., Bass, B., Coffman, R.R., Doshi, H., Dunnett, N., ... Rowe, B. (2007). Green roofs as urban ecosystems: Ecological structures, functions, and services. *BioScience, 57*(10), 823–833. https://doi.org/10.1641/B571005

Places to Grow Act. (2005) SO 2005, c. 13.

Province of British Columbia. (n.d.). Climate Action Charter. https://www2.gov.bc.ca/assets/gov/british-columbians-our-governments/local-governments/planning-land-use/bc_climate_action_charter.pdf

Province of British Columbia, Union of British Columbia Municipalities, Signatory Local Governments. (2007). Climate Action Charter.

Riparian Areas Protection Act S.B.C. 1997 c. 21.

Riparian Areas Regulation B.C. Reg. 376/2004.

Rocky Point Metalcraft Ltd. v. Cowichan Valley (Regional District) BCSC 756, [2012] BCJ No 1043.

Roseland, M. (1992). *Building sustainable communities.* National Roundtable on Environment and Economy.

Rosol, M. (2013). Vancouver's EcoDensity Planning Initiative: A struggle over hegemony? *Urban Studies, 50*(11), 2238–2255.

R. v. Tener, [1985] 1 SCR 533.

Saanich Citizens for a Responsible EDPA. (2015a). [Website no longer available; document on file with author.]

Saanich Citizens for a Responsible EDPA. (2015b). Learn about Saanich EDPA and how it affects you. [Website no longer available; document on file with author.]

Saanich Citizens for a Responsible EDPA. (2015c). Recommended direction for a responsible environmental development permit area (EDPA) in the District of Saanich, 6 November. [Website no longer available; document on file with author.]

Saanich Citizens for a Responsible EDPA. (2015d). Response to FAQ's post on the Saanich website updated June 12 2015. [Website no longer available; document on file with author.]

Sims, A.W., Robinson, C.E., Smart, C.C., Voogt, J.A., Hay, G.J., Lundholm, J.T., ... O'Carroll, D.M. (2016). Retention performance of green roofs in three different climate regions. *Journal of Hydrology*, *542*, 115–124. https://doi.org/10.1016/j.jhydrol.2016.08.055

Slack, E. (2006). *The impact of municipal finance and governance on urban sprawl.* Science Advisory Board of the International Joint Commission.

Smart, A. (2015a, 23 February). Owners irked after Saanich deems properties sensitive ecosystems. *Times Colonist.* http://www.timescolonist.com/owners-irked-after-saanich-deems-properties-sensitive-ecosystems-1.1772291

Smart, A. (2015b, 15 April). Lobby group opposes Saanich environmental protection bylaw. *Times Colonist.* http://www.timescolonist.com/lobby-group-opposes-saanich-environmental-protection-bylaw-1.1823802

Smart Growth Network. (2006). *This is Smart Growth.* https://www.epa.gov/sites/production/files/2014-04/documents/this-is-smart-growth.pdf

Toronto Area Transit Operating Authority v. Dell Holdings, [1992] 1 S.C.R. 32 (SCC).

United Taxi Drivers Fellowship v. Alberta 2004 SCC 19.

Vancouver v Simpson [1976] 3 WWR 97, [1977] 1 SCR 71, 65 DLR (3d) 669, 7 NR 550, 1976 CarswellBC 147.

Washi Beam Holdings Corp. v. West Vancouver (District) [1999] BCJ No 617, 2 MPLR (3d) 118 (SC).

Westfair Foods Ltd. v. Saanich (District) (1997) BCJ No 331, 30 BCLR (3d) 305 (SC), affirmed, [1997] BCJ No 2852, 49 BCLR (3d) 229 (CA).

World Commission on Environment and Development. (1987). *Our common future.* Oxford University Press.

Yung, L., & Belsky, J.M. (2007). Private property rights and community goods: Negotiating landowner cooperation amid changing ownership on the Rocky Mountain front. *Society & Natural Resources*, *20*(8), 689–703. https://doi.org/10.1080/08941920701216586

5 Sustainable Urban Design: The Case of Montreal

HOI L. KONG

Introduction

In 2015, the City of Montreal pursued a consultation process in order to devise a plan to reduce dependence on fossil fuels within its borders. This will be the case study for this chapter, and let me begin by situating the chapter within this book's general treatment of innovation in sustainable municipal design. As we note in the book's introduction, this collection – perhaps most extensively in the Shearmur contribution – sometimes engages directly with academic theories of innovation. But the book's chapters also aim to address innovation in the more colloquial sense of highlighting new developments, in law and other disciplines, that engage sustainability in the municipal context. This chapter takes as its case study Montreal's 2015 novel consultation process and places that new development within legal academic literatures on the city as a commons and sustainable development in cities.[1] But this chapter also can be seen through the lens of the public law literature on regulatory innovation.

Ford (2017, ch. 4) offers a history of legal scholarship on innovation in the regulatory context. For present purposes, the sub-literature on "experimentalism" is particularly relevant, as it provides the legal theoretical background on innovation against which this chapter can be viewed.

---

1 In Canadian law, the term "city" typically denotes a legal entity, whose powers are delegated to it by a province and that in Canadian constitution law would be characterized as a municipality. This legal usage is not necessarily coextensive with how the term is used colloquially or in non-legal literatures. For instance, commentators in the popular press and in academic disciplines, including political science, sometimes will refer to a metropolitan area, whose boundaries are not coextensive with those of the entity to whom a province delegates formal legal powers, as a "city." For this distinction, see Kong (2012).

As Ford notes, Charles Sabel and his collaborators assume that regulated activities will necessarily change and evolve and that therefore no central regulator in any given regulatory area will alone be able to "produce sufficiently informed or flexible responses to local problems" (p. 95). Ford then lists key features of experimentalist regulatory regimes, including "a broadly participatory, non-hierarchical process" and "established processes for ensuring continuous feedback, reporting and monitoring as a means of continually revising and improving practice and results" (p. 95).

Like other legal theories of innovation, including "meta-regulation" and "new governance" theory, experimentalism is process oriented (p. 94). This focus on process also resonates with Margaret Hagan's work on legal design (n.d.). Hagan expressly incorporates "design thinking" into her legal design processes. In particular, she incorporates user-focused and iterative processes into her legal institutional design work. These aspects of design thinking can also be seen in writing on regional social innovation projects undertaken by Espiau and others. These projects create "social platforms" that directly engage local communities and stress collaborative and iterative approaches to regional development (Espiau et al., 2019).

For some legal scholars, including Sabel and other experimentalists, focusing on participatory processes has normative significance: these processes facilitate democratic deliberation, and do so under conditions of empirical and normative uncertainty (see Dorf & Sabel, 1998). As we shall see, this chapter addresses the deliberative democratic implications of the legal literatures on the city as a commons and urban sustainable development. And it argues that the Montreal case study exemplifies innovative design processes that facilitate broad-based democratic participation, in the deeply contested policy area of sustainable development. The objective of this chapter is therefore not to address the *regulatory outcomes* of the process. It is rather to show how the process responds to normative concerns about democratic governance – concerns that are shared by other theorists of public law innovation.

Let me turn now to the two related bodies of writing on municipal law that this chapter will draw upon to address the question of what theories and principles should guide the design of consultation processes about sustainable urban development issues. The first conceives of the city as a commons; the second addresses the challenge of sustainable development in cities. Both are relevant to the process design question that I have identified.

There is a significant common interest in preventing resources from being depleted by the self-interested activities of individual actors. Some of these resources are social. Think, for instance, of the social value of interactions among community members in a park or a neighbourhood street, and how unregulated overuse of either spaces can degrade and exhaust that social good (Foster & Iaione, 2018). Other resources in a city, such as land, can be degraded by unregulated and self-interested uses in ways that harm the environment. On this point, the city as a commons literature overlaps with writing on sustainable urban development (for this overlap, see p. 281) and especially with the latter's concerns about how to develop land in ways that are economically, socially, and environmentally sustainable.

In exploring this chapter's case study through the city as a commons and sustainable urban development literatures, I will *apply* key insights of each body of writing, *identify* their points of overlap, and *critique* these bodies of writing in light of deliberative democratic legal theory. We shall soon see in greater detail that this branch of legal theory focuses on rendering public processes legitimate by obliging public authorities to publicly justify their actions in terms that citizens could reasonably be expected to accept. Such governance, we shall see, respects the rational agency of citizens and does not expose them to state action that merely reflects the will of the majority or of special interests (Gutmann & Thompson, 1996, p. 12).[2] We shall see that this kind of deliberative democratic legal framework provides a response to a serious governance critique of the writing on the city as a commons and on sustainable urban development.

In Part I, I will outline key insights of the literatures on the city as a commons and sustainable urban development, and I will highlight points of overlap between them. In Part II, I will raise criticisms of these bodies of writing, and I will demonstrate how the case study illustrates key aspects of these literatures and provides a deliberative democratic response to the criticisms. The article therefore makes three broad contributions to the relevant literatures: (1) it highlights points of overlap that have not been previously identified; (2) it applies the theories (and these overlaps) to a case study that has not previously been theorized in these terms; and (3) it articulates a critique of the theories that is grounded in deliberative democratic legal theory, and develops a response that draws on the resources of this theory, as well as specific elements of the case study.

Part I: The City as a Commons and Sustainable Urban Development – Insights and Overlaps

The scenario in Montreal identified at the very outset of this chapter gives rise to governance challenges that directly engage concerns about how to develop cities in ways that are sustainable. The legal writing on

---

2 According to the authors, "The basic premise of reciprocity is that citizens owe one another justifications for the institutions, laws, and public policies that collectively bind them" (Gutmann & Thompson, 2004, p. 133).

the city as a commons and on sustainable urban development provide analytical frameworks for addressing these kinds of challenges, and in this Part, I will highlight salient elements of each framework and elaborate on points of overlap between them. These insights set the stage for the discussion of the case study and the prescriptive arguments in Part II.

The City as a Commons

Foster and Iaione have offered the most extensive treatment of the legal concept of the city as a commons. For present purposes, the notion finds roots in theorizing about property law. The authors write, "[L]egal scholars have taken as the starting point the idea that the commons is an unrestricted and unregulated open access resource which allows uncoordinated actors to overconsume or overexploit the resource and then discuss solutions to avoid these tragic outcomes" (Foster & Iaione, 2016, p. 287). Foster and Iaione present city space itself as an example of this kind of commons.

According to them, the unregulated and self-interested use of urban land carries risks of "[c]ongestion and overconsumption ... [which] can quickly result in rivalrous conditions in which one person's use of that space subtracts from the benefits of that space for others" (p. 292). Consider urban parks, which if insufficiently regulated can give rise to what the authors call "'chronic street nuisances' – such as excessive loitering, aggressive panhandling, graffiti, or littering – that eventually begin to rival, if not overwhelm, other users and uses of open spaces" (p. 299).

What is true of specific sites within a city is also true of the city as a whole. Foster and Iaione argue that the unregulated and unrestrained consumption of urban land will result in its exhaustion and can give rise to rivalrous uses – such as the classic example of an industrial plant that is sited next to a residential home – across the territory of the city (p. 295). This conception of the city as a commons describes all the land within the territory of a city as a commons that is susceptible to tragedy.[3]

Foster and Iaione argue that a particular theory of governance can be applied to this governance challenge. Drawing on the work of Elinor Ostrom, Foster and Iaione argue that the city as a commons should be managed in ways that are "neither exclusively public nor exclusively private" (Foster & Iaione, 2016, p. 289). In this view of urban governance, public authorities do not hold a "monopoly" position in regulation, but rather "enable" decision-making by "co-partners or co-collaborators" (p. 290). An example illustrates this approach.

---

3 The authors also set out a site-specific version of the theory. I will omit discussion of it, as it is not relevant to the present case study.

Consider a city government that has adopted a broadly applicable collaborative approach to governing the urban commons. Bologna has developed a "co-city" protocol, which is a "policy and regulatory framework re-shaping the relationship between inhabitants and local administration with regards to urban resources and services" (Foster & Iaione, 2016, p. 348). One element of this framework is a regulation governing "collaboration agreements." The regulation defines "urban commons" broadly[4] and requires the city to identify "spaces, buildings or digital infrastructures which could be the target of actions of care and regeneration, specifying the goals to be pursued through collaboration with active citizens" (p. 347, n263, quoting Bologna's Regulation, § 10[6]). According to the regulation, collaboration agreements identify the "objects of care, and the rules and conditions of collaboration among any group of citizens and the local government or other actors" (p. 347).[5] Furthermore, the regulation provides norms and guidance to collaborators and specifies that the city will transfer technical and monetary support to any collaboration (p. 347).

According to Foster and Iaione, this example of collaborative governance reflects a desire on the part of citizens and non-state organizations to "be part of the decision-making process shaping the lives of city inhabitants" (Foster & Iaione, 2016, p. 347). Foster and Iaione's writing on the city as a commons, therefore, has the potential to provide insights into the kinds of *processes* that the City of Montreal has adopted to address the urban commons governance challenge of reducing dependence on fossil fuels. In the next section, I will turn from this consideration of *processes* to examine the *substance* of the concerns that arise from the case study.

Sustainable Urban Development: Definitions and Instruments

The principle of sustainability – that development that aims to meet present needs should not compromise the environmental conditions necessary for future generations to meet their needs – has been articulated

---

4 The definition of the urban commons is as follows: "[T]he goods, tangible, intangible and digital, that citizens and the Administration, also through participative and deliberative procedures, recognize to be functional to the individual and collective wellbeing, activating consequently towards them, pursuant to article 118, par. 4, of the Italian Constitution, to share the responsibility with the Administration of their care or regeneration in order to improve the collective enjoyment" (Comune Di Bologna, § 2[a]).

5 Actors with whom the city may collaborate include "social innovators, local entrepreneurs, civil society organizations, and knowledge institutions willing to work in the general interest" (p. 348).

in international and domestic legal documents (see Brundtland, 1987, ch. 2; Norton, 2005; Hollings, 2000; *Sustainable Fisheries Act*, 1996; for further discussion, see Kong & Wroth, 2015, p. 1). This principle further emphasizes that development is sustainable when it satisfies economic, social, and environmental objectives, and scholars have argued that to be considered sustainable, decision-making should integrate consideration of these goals (Dernbach, 2003).

Land-use-law scholars have long argued that urban development patterns in North America have caused significant environmental harms that in turn have compromised the capacity of future generations to meet their needs. Scholars have noted that the typical North American city's zoning regime creates urban sprawl, which damages vulnerable ecological systems and contributes to substantial energy consumption and greenhouse gas emissions as a result of, among other things, increased dependence on automobiles for commuting (for discussion, see Beatley & Manning, 1997, p. 8; for fuller discussion, see Kong, 2012a). One example of the environmental harm that results from urban sprawl and poses a risk to future generations is decreasing air quality, which represents a significant threat to the well-being of city dwellers and, if unchecked, will become increasingly serious.[6]

Land use scholars in the legal academy have examined regulatory instruments that can reduce environmental harms caused by dominant patterns of urban development in North America (see, e.g., Salkin, 2010, p. 2; Nolon, 2009, p. 9). As is consistent with the conception of sustainable development set out above, these urban development strategies aim to balance social, economic, and ecological goals. In this vein, Dernbach and Bernstein have argued for sustainable urban growth policies, which aim to "minimize sprawl and maximize sound development opportunities to conserve important lands, preserve the natural environment, protect air and water quality, promote affordable housing through compact development and urban renewal and encourage 'infill' rather than rural development" (Dernbach & Bernstein, 2003, p. 511).

Consider, finally, contemporary forms of regulation that aim to address environmental challenges in general and in land use in particular. Public law scholars writing on environmental questions have come to realize that they typically involve a range of interests and that no single regulatory instrument, whether it be market-based (e.g., an emissions trading regime) or command and control (e.g., detailed and imperative regulation) can adequately respond to this complexity (for a full discussion of

---

6 For a description of the concern and alternative approaches to traditional zoning that attempt to address, see Nolon (2015, pp. 10226–10227).

these ideas and an overview of the relevant literatures, see Kong, 2012a). An example of sustainable land use regulation that evinced these features of environmental regulation has been analysed in the legal literature.[7] The object of the relevant regulation was an abandoned dockyard in Victoria, British Columbia, which gave rise to concerns about the economic viability and environmental sustainability of the proposed development, as well as worries about the effect of development on the stock of affordable housing in the area. In response to this complexity, the City blended together a variety of regulatory instruments, including a bidding process, a specific form of zoning instrument, design guidelines, annual reports, and monitoring, in collaboration with a community organization.[8] In its use of multiple regulatory instruments and convening of various policy actors, this form of regulation has much in common with other complex instruments of environmental governance[9] and with the collaborative governance arrangements that we have seen in Foster and Iaione's discussion of the urban commons.

The City as a Commons and Sustainable Urban Development: Overlapping Frameworks

Now that we have examined key insights of the city as a commons and sustainable urban development literatures, we are in a position to identify three points of overlap. First, we saw at the outset of the discussion of the city as a commons that urban land can be conceived of as a resource that is subject to a "tragedy of the commons." If we view this tragedy through the lens of sustainable urban development, we can specify negative environmental outcomes that are related to land use and that are associated with particular failures to adequately regulate the urban commons. Recall, for an example, the discussion of the degradation in air quality that is linked to land use regulation that encourages sprawl, and therefore increased dependence on automobiles for commuting.

A second point of overlap arises in the role of public engagement in the context of the urban commons and sustainable urban development. We saw that according to Foster and Iaione, community members have an interest in participating in equitable and inclusive urban commons regulation (Foster & Iaione, 2016, p. 335). Citizen engagement is also important for sustainable land use regulation. Policies in this context

---

7 Kong (2012a).
8 *Ibid* at 568–568.
9 See the discussion Kong (2012a).

that incorporate citizen input enable governments to resolve conflicts and build public trust. Christopher Hawkins and Xiao Hu Wang note that local government decisions involving sustainable development are inevitably distributive, create winners and losers, and implicate "concerns over growth." The authors therefore conclude that "participatory planning processes that encourage communication and deliberation provide opportunities to resolve conflicts among stakeholders" (Hawkins & Wang, 2012, p. 13; citations omitted). Hawkins and Wang further argue that in this context, policy development can be impeded by citizens' lack of understanding of the issues. In order to overcome this kind of resistance, planning processes aim to "improve learning and foster education on the importance of sustainability for all parties involved in decision-making" (p. 14; for a fuller discussion, see Kong et al., 2017). Providing opportunities for citizen involvement and engagement is one way of developing this kind of understanding and social trust.

A final point of overlap is a *form* of regulation that is common to both bodies of writing. We saw that both the city as a commons and the sustainable urban development literatures prescribe forms of regulation that depart from simple command and control instruments. To broach this issue of *institutional form* it is helpful to cite a distinction that Foster and Iaione draw between government and governance: "[G]overnance is not just 'what governments do' because governance is not a function limited to the State; rather myriad non-governmental organizations, local neighbourhood associations, individual property owners, etc. can (and already do) play an important role in governing resources" (Foster & Iaione, 2016, p. 333; citations omitted). This insight is similar to one advanced by Ostrom, Tiebout, and Warren (1961), who distinguish service provision from service production. They argue that once a local government has decided to regulate in order to *provide* a given service, it has choices about how to *produce* it: "[I]t can choose to undertake production itself by creating an in-house production unit; it can contract with a private or public actor for production; or it can regulate private activity to achieve public ends" (Kong, 2012b, p. 502; citations omitted).

These distinctions between governance and government, and between service provision and production were implicit in the above descriptions of regulatory instruments. We saw in the discussion of Bologna's co-city initiative that where a municipality decides to regulate in order to preserve an urban commons resource, it can enter into collaborative agreements with citizens and other interested actors, including local entrepreneurs and civil society organizations. And Victoria used a variety of means to regulate a complex environmental matter. In each of these cases, the city did not employ a simple form of command and control regulation. Instead, each city engaged in collaborative governance and used blended institutional forms.

In this Part, I have drawn insights from the writing on the city as a commons and on sustainable urban development, and I have highlighted points of overlap between the literatures. We have seen that unsustainable urban development can create a tragedy of the urban commons, and we have noted that there are good reasons for cities seeking to regulate issues relating to the urban commons and sustainable urban development to engage the public. Finally, we have remarked that the bodies of writing share a common institutional framework that emphasizes collaborative and hybrid forms of governance. With this theoretical background established, we are in a position to critique the theories and apply them, in light of this chapter's case study.

Part II: Theoretical Critique, Application, and Response – The Montreal Case Study

In this Part, I will pursue three objectives. First, I will present a critique of the literatures above that is grounded in deliberative democratic theory. Second, I will set out the essential elements of this chapter's case study. Third, I will conclude by showing how the case study exemplifies aspects of the city as a commons and sustainable urban development literatures and responds to critiques of them in a way that is consistent with deliberative democratic theory.

A Deliberative Democratic Criticism of Collaborative Governance

In the next section, we will see in detail the significance and applicability to the case study of the analytical frameworks of the city as a commons and sustainable urban development, both individually and at their points of overlap. Before turning to apply the frameworks, I will canvass two significant critiques of collaborative governance, which, as we have just seen, is an important part of the analytical frameworks in question. Foster and Iaione signal one aspect of the critique when they write that this form of governance "can present problems of accountability and legitimacy because the decision-making process takes place in settings that bypass representative channels of democracy" (Foster & Iaione, 2016, p. 339). The authors note that there is an additional concern about whether collaborative forms of governance are "able to guarantee equal access by underrepresented groups who are too often unable to access political and larger decision making-processes" (p. 340). In what follows,

I will develop this criticism, using the lens of deliberative democratic legal theory.

Let us begin by defining legitimacy, in municipal law. Legal theorists writing in the deliberative democratic tradition argue that legitimacy in public law (of which municipal law is a subset) demands that state action rest on reasons that can be publicly justified to citizens who are subject to it. In this view, state action is not legitimate when state action expresses the mere will or preferences of majorities or particular interest groups.[10] A state that deprives its citizens of a reasonable justification for actions that affect their interests fails to respect their status as reasoning members of the political community.[11]

An extensive literature describes factors that increase the risk that such regulation will be enacted at the municipal level of government. In part because of their size, the populations of municipalities are relatively homogenous and therefore prone to ignorance of or hostility towards the interests of minorities. Also, in part because of the policy dynamics of zoning, municipal governments are susceptible to interest group capture, in which interest groups dominate zoning regulation for their own purposes.[12] These tendencies towards majoritarianism and capture mean that even representative local governments will likely regulate in ways that will not be justified in terms that those subject to state action would reasonably be able to accept. This problem is exacerbated in circumstances, such as those of collaborative governance, where actors who are not subject to democratic controls take significant part in decisions through which the state, whether through delegation, agreement, or some other means, affects the interests of citizens. In collaborative governance regimes, therefore, the decision-making may be particularly opaque, and some actors participating in the decision-making may not be accountable to the populace, directly or indirectly, through elections. (I say "directly or indirectly" because in the modern administrative state, the elected branches of government often delegate authority to expert, unelected bodies [Kong, 2010a, p. 418].)

---

10 See on this point, Gutmann and Thompson (1996, p. 134): "The principle of publicity requires that reason-giving be public in order that it be mutually justifiable. The principle of accountability specifies that officials who make decisions on behalf of other people, whether or not they are electoral constituents, should be accountable to those people."
11 I have argued for this position in Kong (2010a, 2010b), Kong et al. (2017), Farina et al. (2014).
12 For a summary of this literature, see Kong (2012b).

One way of controlling the risk of majoritarian or special interest domination in the land use law context is to create mechanisms of public consultation. Municipal law statutes in Canada and Quebec impose publicity and consultation requirements on municipal councils when they pass zoning by-laws. The details of these regimes need not detain us here. What is significant for present purposes is that they impose on municipal authorities an obligation to engage with citizens in open forums, and to frame the regulation in terms that are set by legislation. This legislation limits, therefore, the considerations that can be used in justifying a council decision and moreover provide mechanisms for reviewing decisions (Kong, 2010a, pp. 419–420). Such constraints compel municipal authorities to justify their decisions publicly – whether they are taken solely by them or in collaboration with other actors – and limit therefore the extent to which such decisions can be instruments of majority will and special interest capture.

Yet these kinds of mechanisms for public consultation can yield the concerns about equality and inclusion to which Foster and Iaione allude. A significant body of literature suggests that certain groups are over-represented in consultation processes; those who have sufficient time and resources tend to participate most often and most effectively. By contrast, those whose personal or work circumstances prevent them from attending meetings or whose limited resources constrain their ability to pay, for example, child care services so that they can attend, are unrepresented in public meetings. Moreover, among those who do attend, some who do not feel comfortable expressing themselves in the language of the professional middle class will defer to participants who are so at ease, and will therefore under-participate. (For a discussion and for references to the relevant literatures, see Kong, 2010a, p. 422.)

We saw above that one benefit of collaborative governance in the context of the urban commons and sustainable urban development is that it expands the circle of relevant actors in governance. The institutional design challenge for this form of governance lies in ensuring that this expanded circle responds to the concerns about legitimacy and inclusion canvassed above. With these concerns, and with the general outlines of the city as a commons and sustainable urban development literatures in view, we can now turn our attention to the Montreal case study.

The Case Study: Reducing Montreal's Dependence on Fossil Fuels

A public consultation, entitled "Reduced Dependence on Fossil Fuels in Montreal," was initiated in 2015 in response to a petition from a coalition of citizens. In anticipation of the United Nations Conference on

Climate Change (COP 21), the group wanted to provide broadly sourced recommendations to city officials on how to reduce the use of fossil fuels in Montreal (see the timeline in OCPM, 2015). In the paragraphs that follow, I will discuss the case study in some detail. I will then show how it exemplifies aspects of the writing on the city as a commons and on sustainable urban development, and provides responses to the above critiques of those literatures.

The public consultation on reducing Montreal's dependence on fossil fuels was managed by the Office de consultation publique de Montréal (the City of Montreal's Public Consultation Office, or OCPM), an independent agency of the City of Montreal that is charged with (among other things), holding public consultations.[13] By-laws give citizens the right to initiate public consultations on matters falling within the jurisdiction of the City of Montreal and limit the number that can be held in a given year. The by-laws set out the procedures governing the form of a draft petition (Ville de Montréal, 2005, s. 6). If a draft provision meets the specified requirements of form and substance, it is admissible. Within a specified time period after filing, the relevant authorities as well as the petition's designated contact person are to be informed of this fact, and the petition is tabled at the next meeting of the City's executive committee (s. 9). Notice of the beginning of the 90-day signing period is then published on the relevant municipal website and in at least one local newspaper (s. 10). If the requirements governing the form of the petition, the eligibility of signatories, and the threshold number of signatures are met, and the petition is filed in the relevant time period, the public consultation must be held in a reasonable time, and the timetable for the public consultation, as well as background information about it, must be posted on the City website (Part IV). The OCPM and a designated committee of the city council are then responsible for holding the consultation and within 90 days of the end of the consultation must publish a report that reviews the opinions expressed during the process and that "formulates conclusions, opinions or recommendations." (s. 21). The report has "no decisional character" and city council must inform "the people concerned of the results of the public consultation and, as the case may be, of the decisions taken and their grounds" (s. 22).

The consultation on reducing dependence on fossil fuels followed this basic structure and included significant innovations that were aimed at

---

13 Public consultations are typically held on "any draft by-law revising the city's planning program" or "at the request of the city council or the executive committee on any project designated by the council or the executive committee" (*Charter of the Ville de Montreal*, s. 83).

engaging community partners. For example, self-organized groups of citizens could register profiles on the consultation website, which provided background materials. Once registered, groups could download from the site do-it-yourself consultation kits, which included a facilitation handbook, information cards setting out possible themes for discussion, blank cards on which to write out challenges and solutions, and posters for summarizing the consultation. The summaries of consultations could be photographed and uploaded onto the consultation website and shared via the consultation's Facebook and Twitter accounts. In addition, there was an initial information session, at which citizens could ask representatives of the City questions, to which officials subsequently provided written responses. The consultation also included a "hackathon" to which technology partners were invited, as well as traditional public hearings, at which individuals and organizations presented their views. Memorandums were submitted, both by those who presented at the hearings and by individuals or groups who did not. The final report was prepared by the OCPM and included recommendations that were drawn from the consultation materials. Subsequently, the executive committee of the city responded to each recommendation by articulating a position, by identifying policies or actions to be taken, or by providing reasons for declining to take action.[14]

Application of Theories and Responses to Critiques

Montreal's consultation on reducing dependence on fossil fuels reflected key elements of the literatures we surveyed in the previous Part. Consider first the concept of the city as a commons. As we saw, the unregulated use of urban land can give rise to a tragedy of the commons, in which resources are depleted and the well-being of users negatively affected. We saw further that some kinds of common urban resources and negative effects could be environmental, and that future generations could constitute one set of affected users. The consumption of fossil fuels in cities can be understood in these urban commons and sustainable development terms. Iaione has identified city streets as a common resource, subject to degradation. He writes,

> All users undertake and benefit from driving their own vehicles, congesting urban streets and releasing greenhouse gases ("GHGs"), but bear little of the congestion-related and climate related costs of their own driving ...

---

14 The positions were summarized in a document that was made available online by Service de l'environnement (n.d.).

Like other tragedies of the commons, resource users inflict losses upon themselves as a group in terms of the ability to use the resource, by lengthening commute times and degrading the transportation resource. Externalities are also imposed upon non-users, the air-breathing public, in the form of pollution. (Iaione, 2009, p. 891)

The Montreal consultation's background document recognizes these costs of driving in the city and targeted sets of recommendations at them (Ville de Montréal, n.d., pp. 6 and 7). Therefore, in this respect and others, its policy objective of reducing fossil fuel dependence can be understood in terms of the urban commons, and sustainable urban development literatures.

In addition, the consultation directly echoed the importance placed by the urban commons and sustainable development literatures on public engagement. We saw above that Foster and Iaione understand municipal governments to be partners with other actors in the governance of the urban commons, and that in the writing on sustainable urban development, citizen engagement is understood to be important because it builds social trust by informing citizens and helps to address conflicts over resource questions by convening affected stakeholders. We can see these dynamics in play in Montreal's consultation on reducing dependence on fossil fuels. The executive committee of the city, in its response to the recommendations of the consultation report, noted that the process had aimed to "inform and sensitize citizens to the current state of affairs in relation to the consumption of fossil fuels, including identifying those sectors that have the highest levels of greenhouse gas emissions and therefore should be prioritized in the context of the consultation" (OCPM, 2015, p. 1). These aspirations for informing citizens and convening stakeholders in order to address distributive issues could also be seen in the consultation process itself. The OCPM received submissions from a wide range of groups and individuals, including non-governmental organizations, professional associations, industry groups, individual firms, university research networks, religious groups, political parties, government representatives, as well as interested and unaffiliated individual citizens. (For a full list of participants, see OCPM, n.d.).

The consultation can also be seen in light of the writing on the city as a commons and on sustainable development's shared understanding of collaborative governance. We saw above that urban governance in these domains can involve the city regulating in partnership with other actors, and using a variety of regulatory instruments, and not simply deploying mandatory rules. Montreal's consultation itself can be seen as an instance of collaborative governance: the executive committee made policy commitments in light of recommendations that were generated through an extended consultation with a wide range of actors, and some of those commitments did not involve legislative commands. For instance, in addition to pledging to use the city's zoning powers, the executive committee committed to engaging in educational campaigns, to using fiscal tools (including spending, subsidies, and lobbying higher levels of government for financing) in order to pursue sustainable development objectives, and to undertaking public monitoring of its sustainable development policies in order to contribute to a culture of transparency and accountability in the city (OCPM, 2015). These commitments represent blended institutional forms and exemplify collaborative governance.

Thus far in this Part, I have applied insights from the writing on the city as a commons and sustainable urban development to the Montreal consultation, and I hope that in so doing, I have illustrated how these insights play out in a specific context. I will close this Part by answering the objections to collaborative governance and public participation surveyed above.

Recall that collaborative governance in urban settings gave rise to concerns about accountability, because in this form of governance, only the cities, and not their partners, are subject to electoral control. There was a risk, we saw in our discussion of deliberative democratic theory, that this kind of regulation would potentially subject citizens to regulations that reflected the preferences of the majority or private interests, rather than publicly articulated and defensible reasons. The Montreal consultation provides a partial response to this concern. All contributions by private actors to the consultation were made public, and were included in the report and on the OCPM's website. Moreover, the executive committee of the city responded directly to the consultation report's recommendations and gave reasons for either following or not following them. This response to the critique is partial because it may not be available in all instances of collaborative governance in urban settings. But the practices of publishing comments and of governments responding to them by offering publicly available reasons *can* be generalized to other municipal consultation processes. The OCPM process can therefore be seen as a template for some forms of deliberative democratic governance at the municipal level.

Finally, the Montreal consultation suggests a response to concerns about unequal access and participation in consultation processes. As we saw above, the consultation provided do-it-yourself kits so that citizens, in groups of their own choosing, could participate at times and places that were most convenient for them. Moreover, the OCPM provided extensive resources on its site, so that citizens could be informed and participate effectively. By opening access, providing materials to level educational advantage, and allowing participants to select their own groups, these consultation tools respond in part to concerns about traditional municipal consultation processes, which happen at particular times and places, and are susceptible to being dominated by participants who share a particular, privileged profile.[15] Moreover, I would recommend that to the extent the data is available, the OCPM assess whether it has achieved its goal of rendering public consultations more accessible. That information could be published in its annual report.

Conclusion

This chapter has aimed to contribute to the literatures on the city as a commons and sustainable urban development and to thinking about how best to design municipal processes of public consultation. It has done so, by (1) drawing salient points from these literatures, and delineating their overlaps, (2) critiquing these bodies of writing, (3) showing how a concrete instance of public consultation can be understood in light of this writing, and (4) demonstrating how this case study, when viewed through the lens of deliberative democratic theory, can provide specific responses to the critiques and provide potentially generalizable lessons.

I close by considering extensions of the analysis advanced in this chapter. I have focused on the potential for well-designed public consultations to hold governments to account and to reduce the risks of domination. In future work, I aim to examine how institutions such as the OCPM can facilitate deliberation in wider civil society by helping to engrain in a city a culture of deliberation and justification. If the OCPM and its collaborators can help instil in citizens the habits and capacities necessary to engage in deliberation about matters of public significance, they can help to create a broadly shared expectation that decisions affecting the public will be the object of deliberations among decision-makers and affected parties. We may further anticipate private entities engaging in exercises of public consultation and justification, whether they are partners with governments in collaborative governance or not. What I have in mind here is not anything as diffuse as the public sphere that plays a prominent role in, for instance Habermas's early writing on coffeehouse culture (Habermas, 1991). Instead, I envisage structured practices and processes of varying degrees of formality, in a range of settings that could include, for instance, public consultations collaboratively designed and

---

15 Of course, in order to access this open-source material one must have a certain degree of literacy and be able to use the relevant technology. The consultation process therefore reduces but does not eliminate barriers to participation.

administered by municipal governments and private developers. Legal scholars have sometimes argued that in order for laws to be effective, they require a social terrain in which there are engrained habits of thought and practice that make citizens receptive to the laws' demands (Fuller, 1968–1969, p. 536). I suggest that in order for the urban commons and sustainable urban development theorists' latent deliberative democratic aspirations for an accessible, fair, and accountable city to come to fruition, the social terrain of cities should be similarly prepared.

Acknowledgments

I acknowledge with gratitude support for this project provided by the Social Sciences and Humanities Research Council, the McConnell Foundation, and the Centre for Interdisciplinary Research on Montreal. I thank Amy Preston Samson for her outstanding research assistance, Stéphanie Pepin for her work on the citations, and Jeff Kennedy and Sarah Berger Richardson for their challenging questions in a summer 2017 workshop at McGill University's Faculty of Law. I am also grateful for questions from participants at the International Academic Association of Planning, Law, and Property Rights 2017 conference, held at Hong Kong University, at a graduate seminar at Ritsumeikan University, at a faculty seminar at Kwansei Gakuin's Faculty of Law, and at a Cities, Planning, and the Public Interest workshop at the University of British Columbia's Allard School of Law. Finally, I thank Thomas McMorrow for very thoughtful written comments and Professors Takamura (Ritsumeikan University), Kimura (Kwansei Gakuin University), and Stacey (University of British Columbia) for their generosity in hosting me at their respective institutions. Two anonymous reviewers provided careful and thoughtful comments that greatly improved the chapter.

REFERENCES

Beatley, T., & Manning, K. (1997). *The ecology of place: Planning for environment, economy and community*. Island Press.

Brundtland, G. (1987). *Report of the World Commission on Environment and Development: Our common future*, ch. 2, UN Doc. A/42/42.

Charter of the Ville de Montreal, C-11.4.

Comune Di Bologna. Regulation on collaboration between citizens and the city for the care and regeneration of urban commons.

Dernbach, J. C. (2003). Achieving sustainable development: The centrality and multiple facets of integrated decisionmaking. *Indiana Journal of Global Legal Studies, 10*(1), 247–284. http://doi.org/10.2979/GLS.2003.10.1.247

Dernbach, J. C., & Bernstein, S. (2003). Pursuing sustainable communities: Looking back, looking forward. *The Urban Lawyer, 35*(3), 495–532. https://papers.ssrn.com/sol3/papers.cfm?abstract_id=983502

Dorf, M. C., & Sabel, C. F. (1998). A constitution of democratic experimentalism. *Columbia Law Review, 98*(2), 267–473.

Espiau, G., Duong, P., Moreno, I., & Fisher, J. (2019). New peacebuilding and socio-economic development approaches in Asia. https://undp-ric.medium.com/new-peacebuilding-and-socio-economic-development-approaches-in-asia-a31e715567d8?

Farina, C., Kong, H., Blake, C. Newhart, M. J., & Luka, N. (2014). Democratic deliberation in the wild: The McGill Online Design Studio and the Regulation Room Project. *Fordham Urban Law Journal, 41*, 1527–1580.

Ford, C. (2017). *Innovation and the state: Finance, regulation and justice.* Cambridge University Press.

Foster, S. R., & Iaione, C. (2016). The city as a commons. *Yale Law & Policy Review, 34*, 282–349. https://papers.ssrn.com/sol3/papers.cfm?abstract_id=2653084

Fuller, L. L. (1968–1969). Law's precarious hold on life. *Georgia Law Review, 3*, 530–545.

Gutmann, A., & Thompson, D. (1996). *Democracy and disagreement.* Belknap Press.

Gutmann, A., & Thompson, D. (2004). *Why deliberative democracy?* Princeton University Press.

Habermas, J. (1991). *The structural transformation of the public sphere: An inquiry into a category of bourgeois society.* MIT Press.

Hagan, M. (n.d.). *Law by design.* https://www.lawbydesign.co/

Hawkins, C. V., & Wang, X. H. (2012). Sustainable development governance: Citizen participation and support networks in local sustainability initiatives. *Public Works Management & Policy, 17*(1), 7–29. https://doi.org/10.1177%2F1087724X11429045

Hollings, C. S. (2000). Theories for sustainable futures. *Conservation Ecology, 4*(2), 7. https://doi.org/10.5751/ES-00203-040207

Iaione, C. (2009). The tragedy of the commons: Saving cities from choking, calling on citizens to combat climate change. *Fordham Urban Law Journal, 32*(3), 890–951. https://ir.lawnet.fordham.edu/ulj/vol37/iss3/7/

Kong, H. L. (2010a). The deliberative city. *The Windsor Yearbook of Access to Justice, 28*(2), 411–433. http://doi.org/10.22329/wyaj.v28i2.4507

Kong, H. L. (2010b). Something to talk about: Regulation and justification in Canadian municipal law. *Osgoode Hall Law Journal, 48*, 499–541.

Kong, H. L. (2012a). Sustainability and land use regulation in Canada: An instrument choice perspective. *Vermont Journal of Environment Law, 13*(3), 553–574.

Kong, H. L. (2012b). Toward a federal legal theory of the city. *McGill Law Journal, 56*(3), 473–502. https://doi.org/10.7202/1009065ar

Kong, H. L., Luka, N., Cudmore, J., & Dumas, A. (2017). Deliberative democracy and digital urban design in a Canadian city: The case of the McGill Online Design Studio. In C. Prins, C. Cuijpers, P. L. Lindseth, & M. Rosina (Eds.), *Digital democracy in a globalized world* (pp. 180–200). Edward Elgar. https://doi.org/10.4337/9781785363962.00017

Kong, H. L., & Wroth, L. K. (2015). Introduction. In H. L. Kong and L. K. Wroth (Eds.), *NAFTA and sustainable development: History, experience and prospects for reform* (pp. 1–12). Cambridge University Press.

Nolon, J. R. (2009). The land use stabilization wedge strategy: Shifting ground to mitigate climate change. *William & Mary Environmental Law and Policy Review, 34*, 1–54. https://papers.ssrn.com/sol3/papers.cfm?abstract_id=1520629

Norton, B. G. (2005). *Sustainability: A philosophy of adaptive ecosystem management.* University of Chicago Press.

Office de consultation publique de Montréal (OCPM). (n.d.). Déroulement de la consultation. http://ocpm.qc.ca/fr/energies-fossiles/documentation

Office de consultation publique de Montréal (OCPM). (2015, 15 October). Presentation of the consultations. http://ocpm.qc.ca/sites/ocpm.qc.ca/files/document_consultation/2aeng.pdf

Ostrom, V., Tiebout, C. M., & Warren, R. (1961). The organization of government in metropolitan areas: A theoretical inquiry. *American Political Science Review, 55*(4), 831–842. https://doi.org/10.1017/S0003055400125973

Salkin, P. (2010). Sustainable development, climate change and land use for local governments. *New York Zoning Law and Practice Report, 11*, 1–10. https://www.researchgate.net/publication/228157552_Sustainable_Development_Climate_Change_and_Land_Use_for_Local_Governments

Service de l'environnement. (n.d.). Positions du Comité exécutif sur les recommandations du rapport de l'OCPM sur la réduction de la dépendance aux énergies fossiles. Office de consultation publique de Montréal. http://ocpm.qc.ca/sites/ocpm.qc.ca/files/pdf/P80/8.2_positions_ce_ocpm_energies_fossiles_v18.pdf

Sustainable Fisheries Act. (1996). Pub.L. 104–297, 110 Stat. 3565, amending 16 U.S.C. §§ 1851–1861.

Ville de Montréal. 2005. By-Law Amending the By-Law Concerning the Montreal Charter of Rights and Responsibilities, 05-056-01.

Ville de Montréal. (n.d.). Reduced dependence on fossil fuels in Montréal. Office de consultation publique de Montréal. http://ocpm.qc.ca/sites/ocpm.qc.ca/files/document_consultation/3.1_anglais_ocpm_fossil_fuels_en_1.pdf

6 The Implications of Stakeholder Group Involvement in Urban Sustainable Development

ALEXANDRA FLYNN

Introduction

Neighbourhood development is subject to contestation and negotiation by many actors. Each city government has its own processes for negotiating local development, ranging from information dissemination and consultation to full participation. While there is a general commitment to civic engagement, specific processes may not be uniform across the city, and participation is not evenly available to all interested actors. There are particular implications for sustainable development initiatives, as stakeholder groups may have their own views on how "sustainability" fits into the existing neighbourhood identity and to whom benefits should be directed. In addition, particular stakeholder groups may dominate local representation or may have interests in certain kinds of development. In Toronto, stakeholder groups, such as resident and business associations, can play a major role in the development process, representing local interests from project proposal to implementation.

To understand how business and resident organizations overlap in sustainability projects, this chapter examines two vignettes from initiatives undertaken along a downtown corridor in Toronto. Toronto's Bloor Street corridor has significance as a historical home to newcomer populations and is the site of rapid gentrification as Toronto's status as a global city has intensified. Moreover, it serves as an important local transit route connecting the city's east and west sides. Stakeholder groups are well mobilized along this corridor and have been historically connected to civic controversies such as a proposed Spadina Expressway, an initiative that was ultimately abandoned following widespread opposition from local groups.

The chapter first sets out the literature on how business and resident groups affect urban decision-making, including the meanings and tensions of sustainability where neighbourhoods are undergoing rapid development and growth. Second, it outlines the specific place-based considerations of the Bloor corridor, including the overlapping representation of business and resident associations along this corridor. This section then offers two vignettes of recent sustainable development initiatives – the Mirvish redevelopment and the Bloor Street bike lane pilot – to illustrate the formal and informal roles given to stakeholder groups through official city processes. The mixed methodological approach involves document analysis and ethics-approved interview data.

The chapter concludes with the implications for stakeholder representation in sustainable initiatives and opportunities for more inclusive participation in municipal planning projects. First, municipal engagement processes are initiated and executed ad hoc. Second, local advocacy groups serve a crucial role in sustainable development processes, raising questions about the accountability of these groups to their members, local residents, and the city at large. Third, local advocates advanced trade-offs in return for their support of sustainable projects, including amenities such as day cares and park space, and assurances regarding the economy and jobs. As much-needed sustainability initiatives are introduced in Canadian cities generally, and this urban corridor specifically, the role, reactions, and mobilization of stakeholder groups must be understood.

Localized Governance in Urban Neighbourhoods and Tensions in "Sustainability"

In 2011, the United Kingdom introduced the *Localism Act, 2011*, which tried to define what was meant by "local" (Government of the United Kingdom, 2011). In her response to the new legislation, Antonia Layard concluded that the legal construction of local in the UK context jarred with other conceptualizations of the term and the place. "Yet once legally implemented with defined boundaries," she rightly observes, "a locality or neighbourhood takes on a new administrative, political and sometimes socially constructed reality" (Layard, 2012, p. 135). In Toronto, the city's local boundaries and bodies have been created over time, and include wards, community councils, and bodies like resident and business associations. To Layard, law's work in such local institutions has taken on a "constructed reality" (p. 135). Put more simply, neighbourhoods may have their own unique local governance models. This section sets out the justifications for localized governance, the role of stakeholder groups in crafting local decision-making models, and the implications of local governance on notions of sustainability.

Justification for Local Governance

Across interdisciplinary perspectives, localized decision-making is seen as a panacea to fulfil the democratic ideal of representation that is closer to the will of the community. In the 1950s, Jane Jacobs passionately advocated for the importance of neighbourhoods in the built form and the decision-making of urban areas (Jacobs, 1961). In her view, local decision-making was more legitimately democratic and connected to the interests and desires of those within neighbourhoods. Jacobs's influence both theoretically and practically had profound application to Toronto, which became her home in later years, and became the canvas upon which she illustrated the power of stakeholder groups to stop the development of the Spadina Expressway in a then-bohemian section of the city. In their study on whether neighbourhood associations encourage more political participation, Jeffrey Berry, Kent Portney, and Ken Thomson (1993) argue that the "key to making America more participatory may be making political participation more meaningful in the context of the communities people live in" (p. 4). They suggest that collective challenges are best understood in a more narrow geographical space and ultimately lead to decisions that are better for society as a whole. Elena Fagotto and Archon Fung (2006) conclude that such bodies permit deliberative democracy based on increased neighbourhood capacity for collective action and neighbourhood development.

The question, then, is how to best achieve localized democracy. According to Jürgen Habermas, there is an ongoing negotiation between and within groups regarding the boundaries of neighbourhood or community (Habermas, 1962). One way to facilitate negotiation is through public participation. Fung and Erik Olin Wright (2001) encourage governments to increase avenues for public participation, including devolving powers to resident-run associations and giving decision-making power to residents. They suggest that tying public power to participation can create connections between these associations and the overall quality of democratic governance. However, their arguments acknowledge that the balance of power in the municipal setting is disproportionately weighted towards those with greater resources, putting the obligation squarely on these weaker voices to take part in participation exercises.

To many scholars, localized democracy should be set out formally in law in order to ensure fair and participatory governance within the city. Erwin Chemerinsky and Sam Kleiner (2013) tout the benefits of local councils, including that they are uniquely positioned to allow historically marginalized residents to engage in the political life of the city. Gerald Frug advocates for the adoption of charrettes, particularly in planning, which are lengthy negotiation sessions that bring together diverse interests like developers, neighbourhood residents, bankers, and city officials to provide concrete feedback on development projects and to educate people on the costs of zoning policies (Frug, 1980). He argues that these and other community-creating strategies should be firmly embedded in the fabric of local government law, that community can be asserted, not merely facilitated, and that local governments are obliged to do so (p. 1104). The obligation that Frug speaks of is based on the normative view that municipal governments have an obligation to facilitate political participation. In each of these conceptions, the state can and should retain the power to craft the rules for neighbourhood governance (Briffault, 1990, p. 394).

BIAs and Neighbourhood Associations as Micro-Democracies

Like in many other urban centres, business improvement areas (BIAs) and neighbourhood associations are the principle bodies that play a role in local governance in Toronto (Morçöl & Wolf, 2010, p. 908). These small-scale bodies may act as brokers between the public and democratic institutions. In their comprehensive study of the nature of BIA governance, Göktug Morçöl, Triparna Vasavada, and Sohee Kim (2014) noted that BIAs can be conceptualized in three different ways: as tools of governmental policies, as actors in urban governance networks, and as private governments. The study showed that BIA directors play a profoundly important role in this overall question of urban governance, as their involvement in the city governance becomes "deeper and wider" over the years (p. 814). Similarly, in a study of Toronto's Downtown Yonge BIA, researchers observed that the objectives of BIAs tend to evolve from basic operational and tactical tasks to more strategic tasks. This leads to improved data, cost-effective decision support, and increased coordination at the city, regional, provincial, and national levels (p. 802).

Likewise, neighbourhood associations have an impact on local governance. In his exploration of resident associations, Richard Thompson Ford (1999) notes,

> Residence in a municipality or membership in a homeowners association involves more than simply the location of one's domicile; it also involves the right to act as a citizen, to influence the character and direction of a jurisdiction or association through the exercise of the franchise, and to share in public resources. (p. 847)

Chaskin and Greenberg (2015) believe that neighbourhood associations are central to local governance, through fostering collective decision-making and encouraging civic engagement, whether or not they are offered administrative and financial support. Even where neighbourhood associations are not part of formal processes, they are embedded in governance mechanisms by leveraging relationships with allies and partners and negotiating on behalf of their membership (p. 264). In Chicago, they have been able to use this "interstitial" space successfully to shape policy and allocate resources in the public realm, ultimately playing a more direct role in governance (p. 265). In short, neighbourhood associations influence the nature and process of development within local areas (Buckman, 2011).

BIAs and neighbourhood associations are sometimes facilitated by state regulation and are the means by which public power in the local governance context can be seen as more legitimate, as they facilitate the kind of public participation advanced earlier in this chapter (Chemerinksy & Kleiner, 2013; Fung & Wright, 2001). However, while these institutions advance democratic ends, they require consideration of public/private tensions, as well as distributive justice and other effects. Two main concerns have been raised by theorists about BIAs and neighbourhood associations. First, some theorists are concerned that neighbourhood and community associations promote utilitarian objectives. Alexander and Peñalver (2009) believe in a human or political need to belong, to participate, and to contribute. Under this conception, community is not a place; it is a coming together of people. They argue that the territorial conception of community – namely that boundaries create togetherness – has destroyed the conditions under which the intimate relationships that characterize communities may develop. As a result, associations and institutions are the "new" community, where "solidarity is based on transitory convergences of instrumental objectives, with these bodies having replaced community as the dominant mode of group life in modern America" (Alexander & Peñalver, 2009). Chaskin and Garg (1997) suggest viewing the neighbourhood association along a spectrum. At one end of the spectrum, the association is a separate but complementary institution to local government, offering goods and services beyond the scope of local government; yet farther along, it is incorporated into local government as formal methods of representation and action; and at the other end it is in opposition to a local government. As such, neighbourhood associations can be thought of as an umbrella term incorporating organizations that serve widely different roles in localized governance, depending on their unique structures and purposes.

BIAs go even farther in exacerbating the tension between public and private, as they represent private interests, yet are often officially sanctioned by municipal governments. James Wolf (2006) emphatically states that BIAs are "a part of urban governance and public administration" (p. 70). Wolf argues that BIAs must be placed within the public administration context, even if their objectives focus on the "private" concerns of their members (Wolf, 2006). Lewis notes that as BIAs become service providers, development brokers, and place makers, there is a corresponding retreat of municipal government (Lewis, 2010, p. 203). Randy Lippert and Mark Sleiman (2012) suggest that BIAs "defy a public or private designation, encounter resistance and failure, and produce and transfer knowledge for their own and other institutions' purposes" (p. 62). As such, they are not simply private actors seeking additional power, and these complex organizations do not fit easily within particular descriptions as exclusionary or inequality enhancing.

A second concern relates to the representativeness of such groups in relation to the wider neighbourhood population. Even if small-scale decision-making bodies are fundamental to civic participation, there must be a link between these bodies and the public; and these groups must have a link to political decision-making (Thomson, 2001, p. 5). However, neighbourhood associations and BIAs disproportionately allow for the public engagement and influence of economically privileged residents (Levin-Waldman, 2013). Neighbourhood associations, in particular, are seen as dominated by homeowners who are white and middle class, who do not reach out to other members of the communities, and who focus largely on land use rather than social issues (Alarcon De Morris & Leistner, 2009, p. 48). In Washington, BIAs have purportedly contributed to racial and cultural inequality by favouring the views of mostly white property owners in their decision-making (Mallett, 1993). Even amongst their members, BIAs may in fact limit democracy and exclude particular perspectives of public space, reinforcing political dynamics that exclude marginalized and low-income residents, as well as small businesses (p. 405).

Local Governance and Sustainability

There is no easy definition of "sustainability," which has been used in academic materials to mean liveability, quality of life, urban environmental quality, and well-being (Carmona & de Magalhães, 2009, p. 522). The dictionary definition of "sustainability" has a dual meaning that directly touches on the inherent tension in city planning and localized democracy. In one iteration, sustainability refers to "the ability to be maintained at a certain rate or level" (*Merriam-Webster*). However, the term also means "avoidance of the depletion of natural resources in order to maintain an ecological balance" (*Merriam-Webster*). This chapter adopts Julian Agyeman and Tom Evans's definition of sustainability as "the need to ensure a better quality of life for all, now and well into the future, in a just and equitable manner, whilst living within the limits of a supporting ecosystem" (Agyeman & Evans, 2003, p. 36).

An often-used typology for sustainable development in urban planning suggests that sustainability must be based on a balance between social equity, environmental protection, and economic growth (Agyeman & Evans, 2003, p. 37). Social equity includes meaningful participation by a broad range of stakeholders. This typology can operate at both the "city-wide" and "local" scales. As an example of city-wide visioning of an overarching sustainability strategy, a Mayor can offer the leadership necessary to enable relevant business and civic stakeholders to work together with city decision-makers, and by earmarking resources, such as staff, land, and funding (Fuchs, 2012, p. 53). For example, both Toronto and Vancouver have introduced a sustainability model to guide planning in its downtown core, notwithstanding city-wide or regional plans that consider sustainability from a much larger scale (McCann, 2011, p. 108). This model prioritizes cycling and walking paths, a planning focus on the water landscape, and mixed-use zoning and design that encourages use of streets. City-wide or site-specific environmental advocacy groups can also achieve significant sustainable development results where they are given the agency to meaningfully partner with or provided resources by a municipality (Chisholm, 2004).

However, this chapter focuses on a different set of challenges – projects that are meant to have sustainability objectives, but are not city-wide initiatives. These projects remain at the local level, led by neighbourhood-based planners whose aim is to achieve balance amongst the potentially conflictual objectives of social equity, environmental protection, and economic growth (Agyeman & Evans, 2003). The result is what some scholars call a "political quagmire," meaning a combination of

> need for past and future electoral support from the business community; the presence of a ward city council structure leading to geographically based representation of interests and a clear blame structure; upcoming, likely contentious elections; different constituent groups with differing and often competing goals; conflict within groups about goals; and few economic development resources to allocate. (Reese et al., 2001, p. 425)

In this kind of situation, where the city is initiating planning in downtown areas with multiple interested stakeholders,

> serendipity is the operating principle. No single solution, policy, level of government, agency, business, or individual is expected to do it all. The evolving idea is that city centre renewal efforts must be multifaceted, involving the separate actions of the private, public, and non-profit sectors as well as innovative partnerships between business and government. (p. 431)

In such developments, "Organizational fragmentation is admirable, rational planning is less significant, and policy implementation is mostly *ad hoc*" (p. 431).

The two crucial questions for this chapter are, first, Who plays a role in these "innovative partnerships," and second, How are they brought together? First, in social movement literature, city-based activism tends to start at the local level in response to some kind of serious or complex issue (Nicholson et al., 2010, p. 185). Activists may form an organization aimed at influencing government decisions. Coalitions amongst various groups help to strengthen local organizations by offering more financial and human resources, especially in situations where the group must face a large bureaucracy or well-funded business interests (Nicholson et al., 2010). As organizations grow in status and size, they become more professional, with staff and skills that help organizations in achieving their objectives (Nicholson et al., 2010). These organizations have their own views of what sustainability entails that may differ from those of developers and city governments. Second, what role do these organizations play in decision-making processes? The policy area in question can also reveal different participatory processes. For example, land-use planning processes have been modified in many urban centres to empower community members to make decisions with governments and developers (Agyeman & Evans, 2003, p. 48). Transportation planning, including cycling, has not typically included stakeholders in decision-making, leading to more targeted activism by cycling advocates (Agyeman & Evans, 2003).

The next section engages squarely with sustainable development and the role of two principle stakeholder groups – businesses and resident associations – by querying their role in two recent initiatives: the Mirvish redevelopment and the installation of bike lanes. The objective is to understand how these groups were engaged in the projects and the perspectives of sustainability that they brought to the debate. The next section also animates the central questions in this chapter: Which organizations play a role in particular developments and how do they influence the decision-making process? While this chapter does not aim to find a grand theory for neighbourhood participation, it does seek to tell a story of a particular time and place that hopes to be of use in explaining local participation in urban sustainability efforts as critical initiatives are advanced in Canadian cities.

A Portrait of a Neighbourhood: Stakeholder Representation and the "Greening" of Toronto's Bloor Corridor

In Toronto, BIAs and neighbourhood associations use geographically demarcated boundaries to determine and represent their members. Both aim to shape planning decisions and the public realm. At times, the functions of these bodies overlap, with some neighbourhood associations purporting to represent business interests and BIAs permitting residents to sit on their boards. This section examines the overlap of BIAs and neighbourhood associations by first examining what is known about these bodies in Toronto and, next, applying two case studies.

The Overlap of BIAs and Neighbourhood Associations

Despite their similarities as representatives of particular interests within communities, there are important distinctions between resident and business associations. In Toronto, a BIA comprises commercial and industrial property owners and business tenants within a specified geographic area district, which is officially approved by the City to stimulate business and improve economic vitality (City of Toronto, 2016; Hoyt & Gopal-Agge, 2007). BIAs have institutional support from the City of Toronto, including comprehensive information on the city's 81 associations, which are stored in a publicly accessible website; providing training and support to their organizations on their governance; collecting and remitting a levy, which enables BIAs to hire staff and consultants, hold events in their neighbourhoods, and promote local business; and Councillor membership on every board. From a legal perspective, BIAs are bound by the requirements of Toronto's procedural bylaw and must adhere to strict accountability and representation requirements.

By contrast, neighbourhood associations do not have institutional support. Toronto's 184 neighbourhood associations are not officially sanctioned by the city, but are instead private organizations comprising neighbourhood residents who are scattered unevenly across the city (Miller, 2013, p. 145). The only legal requirements to which neighbourhood associations must adhere are those contained in legislation where such bodies are incorporated (Province of Ontario Corporations Act,

1990). Such legislation, however, does not concern any responsibilities of the associations *viz.* the public, including accountability. Their budgets rely on donations and membership fees and are typically limited. One Toronto Councillor described them this way:

> [T]here is no sort of legal or political structure around what [neighbourhood associations] are. And so you have different models: there's the ratepayer association, the residents association, and the community association. Ratepayer being property owners, or residential property owners. Resident being residents, so tenants are included. And community association, which from time to time will take in, say, a representative of a hospital that's in the neighbourhood, or some businesses. So there are no clear boundaries in terms of who is allowed in and who isn't. (City of Toronto Councillor or staff member #3, 2016)

As such, very little is documented regarding the sizes, geographical boundaries, objectives, and sources of revenue of the city's neighbourhood associations. One Councillor affirmed that there is "no sort of legal or political structure around what those things are" (City of Toronto Councillor or staff member #3, 2016). Toronto has hundreds of neighbourhood associations, which differ dramatically in their size, structure, formality, history, and involvement in local governance. Neighbourhood associations are not formally embedded within the City's bureaucratic structure, therefore, data collection is difficult, as neighbourhood associations come and go and the organizations generally have limited resources. Unlike New York's community committees or Los Angeles's neighbourhood councils, there are no departments within the City of Toronto responsible to assist or otherwise keep track of neighbourhood associations (City of New York, 2017; City of Los Angeles, 2017). Overall, there is an absence of clear information on elections and other internal practices that could shed light on their adherence to principles of democracy and meaningful representation, and their commitment to accountability practices.

However, for members of staff there may not be significant differences between these bodies in their local advocacy roles. A senior staff planner told me, "They're stakeholder groups. I wouldn't weigh one over the other necessarily. I'm very interested in making sure that we hear from them both" (City of Toronto Councillor or staff member #4, 2016). The importance of having both kinds of associations is connected to the role played by city Councillors in helping to form and further the involvement of BIAs and neighbourhood associations in their wards, as "the glue between different neighbourhoods" (City of Toronto Councillor or staff member #2, 2016), bodies that allow Councillors to have "greater reach within a community" (City of Toronto Councillor or staff member w #1, 2016), and that serve as "citizen experts" (City of Toronto Councillor or staff member #3, 2016). It is not surprising, then, that the boundaries of the associations overlap, whereby multiple neighbourhood associations and BIAs may claim to represent a particular area (Topping, 2015).

A senior planning staff member explained that the planning model in Toronto is like "a spoke in the wheel," and once the application is received from the applicant, "we send it out to everybody, they give us their comments, we circulate it out, they give us their comments back, we go out, we come back, we go out, we come back" (City of Toronto Councillor or staff member #4, 2016). The role of the planner is to consider these interests among the medley of other voices.

> I call them citizen planners, but they're people who know stuff that we don't know ... People know the city very well and I'm very, highly respectful of the way the people experience their neighbourhoods on a daily basis and the knowledge. Everybody's got their biases and everybody's got their proclivities and, you know, some of them are going to be more informative and some are going to be less ... [But] these ideas that are out there are all worth scrutiny. (City of Toronto Councillor or staff member #4, 2016)

The planning process includes "all the inputs that we have into the process, both professional, technical inputs that come, legal inputs, awareness of the law, legislative input, and agencies, stakeholders, and the public" (City of Toronto Councillor or staff member #4, 2016). Robust stakeholder engagement and consultation

> tend to follow kind of like a typical map, you know a process map, where once, oftentimes before somebody files an application there'll be some consultation happening, if they're smart. They'll be engaging their neighbours, they'll be engaging the community ... Then after the application comes in we have more formalized consultations, sometimes the Councillors set up working committees, working groups, where you really get into the nitty-gritty and the detail. (City of Toronto Councillor or staff member #4, 2016)

This process leads to revision based on what the public "said we got right, what they said we got wrong," that is complemented by the professional opinions of staff and the degree to which the applicant is willing to modify the project. Ultimately, staff write a report to community council to introduce a by-law and then say, "Over to you, decision makers." Even then, there is "endless talk," where City Council then makes a decision, and the developer or public may appeal.

While BIAs are, as one Councillor called them, bodies that "create their own little tiny tax base and they tax and spend on themselves, and they all act in self-interest" (City of Toronto Councillor or staff member #2, 2016), neighbourhood associations are more likely to react to local politics and determine their activities accordingly, rather than have stand-alone missions separate from what the city does (Portney & Berry, 1997, p. 634). What remains uncertain is the extent to which BIAs and neighbourhood associations cooperate with one another. In his study of the development of the "creative city" in one of Toronto's largest BIAs, the Entertainment District BIA, Sébastien Darchen observed the tension between the local planning process, in which the BIA carried a far stronger voice in community deliberations and their interests were specific to the advantages for the member businesses (Darchen, 2013, p. 197). While Council ultimately supported a mixed-use neighbourhood – which includes a more diverse range of economic activities than those proposed by the BIA, including an environment where emerging artists can live and work in the neighbourhood – Darchen concluded that the promotion of arts and culture as imagined by the BIA led to revitalization of the area (p. 201). Building on Darchen's work, the vignettes that follow examine the overlap between BIAs and neighbourhood associations by asking whether and how these bodies influence decision-making in two sustainable development initiatives.

The following case studies illustrate how the normative and empirical questions identified in the previous sections, including participatory processes, the roles of advocacy groups, and accountability to the public play out in two initiatives involving BIAs and neighbourhood associations, and how the specific interests of these groups were embedded in larger networks of state authority and political and economic power.

Mirvish Redevelopment

In 2010, "Honest" Ed Mirvish, a local businessman, died. In 2013, his family sold the iconic Honest Ed department store and the nearby Mirvish Village on the corner of bustling Bathurst and Bloor. The land was purchased by Westbank, a Vancouver-based developer, with the intention of tearing down the enormous dollar store with the heritage-valued sign and constructing a large, mixed-use neighbourhood in its place, with sustainability as a pillar in the proposed development (Novakovic, 2016).

Community consultation began with neighbourhood associations before the development proposal went to City Planning. The four local associations saw themselves as important participants in shaping the development process, forming a task group to oversee consultation with the City (Seaton Village Residents' Association, 2014). One of their first tasks was to conduct a survey on the development of Honest Ed's and Mirvish Village. They noted,

> [I]f we are to play an active and positive role in shaping the future of our neighbourhood, we need to articulate the values that make this neighbourhood an attractive and vibrant place in which to live. A vision for the neighbourhood is an essential basis for negotiations with the city and developers, however large or small. Furthermore, the proposed changes in the municipal approval process for development assume a neighbourhood vision that reflects the collective values of the community. (Seaton Village Residents' Association, 2014)

The neighbourhood collective had praise for the developer and architect, who expressed

> a strong foundation of community consultation and of mindfulness towards stakeholder and resident's/neighbourhood concerns, priorities, and values INCLUDING the deep and varied feelings that exist for Honest Ed's and recognizing the historical significance and emotional attachment so many who live in or visit the neighbourhood have for Mirvish Village. (Seaton Village Residents' Association, 2014)

The developer and architect had previously worked together in Vancouver to redevelop the Woodward's building into a mixed residential-business area on the grounds of a former shopping centre. As the developer's website notes,

> Like Honest Ed's, Woodward's was once a department store, and fragments of the century-old building were incorporated into the redevelopment, along with homages to the store's distinctive landmark sign. It took [the developer and architect] six years to complete, and included several rounds of community consultation aimed at developing a project that was sensitive to the needs of all of its stakeholders. (Brissenden, 2015)

In 2015, the City of Toronto's Planning Division received the formal application to amend the Official Plan and Zoning By-laws to redevelop a number of properties bordered by Bloor Street West, Bathurst Street, Lennox Street, and Markham Street, as well as a number of properties on the west side of Markham Street, known locally as the site of Honest

Ed's and Mirvish Village (Director Community Planning, 2017). The development proposal received by the City would result in an increase in building height from 5 to 26 storeys. The development would include over 800 new rental residential units, a new public park, a day care facility, and public realm improvements on Markham Street. The changes to the neighbourhood would require an Official Plan Amendment, but were in conformity with the requirements of the Provincial Policy Statement, providing for an appropriate range of housing types and densities to meet projected requirements of current and future residents (pp. 21–27). Despite the initial positive start, once the application proposal was received, neighbourhood associations were very concerned about the proposed development. The well-established Harbord Village Residents' Association (HVRA), in response to a 5 October 2015 staff report, lamented the lack of consultation with resident groups and proposing a "bottom-up" approach that would consider the needs of existing residents (Toronto and East York Community Council, 2015). These needs included housing for a full income/age range, attention to new infrastructure, and reduced density (Director Community Planning, 2017). The final deal included a contribution of $4 million for affordable rental housing (Director Community Planning, 2017).

In response, the City Planning Department undertook a wide variety of consultations between 2015 and 2017 on the Mirvish redevelopment, as outlined in the first staff report on the matter (Toronto and East York Community Council, 2015). They included several "large format meetings," which were drop-in style meetings with City staff and the local Councillors that over 650 people attended (p. 18). City Planning staff also met directly with local neighbourhood associations and other organizations (p. 19). A number of residents formed the Mirvish Village Task Group, providing written correspondence to City staff, gathering and disseminating information to the public (p. 18). However, in this case, City Planning staff also opted for a far more enhanced engagement strategy, perhaps owing to the discord expressed by the neighbourhood associations. City staff convened a "Planning Discussion Group" in October 2015 shortly after the community council meeting, with residents, business owners, and local representatives to "contribute local experience and observation to City staff's review of the proposal and to help inform staff comments and eventual recommendations" (p. 18). To City staff, such a group is invaluable. As one senior staff member explained,

> Instead of just going constantly out to big events, you have a small group that stays with you all the way through and they're like your touchstone, where you come back to them and say, ... "Can we run our draft presentation by

you, how is it reading, ... is it making sense to you?" ... [I]t's a place for us to go to get more intimate feedback and guidance. (City of Toronto Councillor or staff member #4, 2016)

As in the case of the Mirvish Village, the members of the advisory group are "sometimes self-appointed," or the Councillors asked the residents association to nominate a representative so that there was a distribution of feedback of different stakeholders in the area, which the staff member considered to be "an even more sophisticated model." This model also conforms to Jack Meek's hypothesis that expanded public participation with bureaucratic and political support creates a "stronger bond" for enabling urban environmental solutions (Meek, 2008, p. 420).

The group included members from the four neighbourhood associations who had launched the dedicated advocacy group, the local BIA, local business owners and residents unaffiliated with the formal groups, and two local Councillors. To a senior staff member, the Mirvish Village consultations is a "pretty advanced, sophisticated level of engagement" and "not everybody does this, not every developer." This process included open houses, topic specific discussions, and "a real transparency to the whole thing" (City of Toronto Councillor or staff member #4, 2016). The Planning Discussion Group met eight times from their formation over the next eighteen months. The meetings were structured around themes like Heritage Preservation and Transportation, and attended by City staff who specialized in that meeting's specific theme. The group met twice to discuss the draft conclusions and policies of planning studies and the Official Plan Amendment. The group also met with the developer to preview work on the first set of revisions to the proposal and to discuss the format for the City's second large format consultation meeting. The City Staff report noted that, "Should City Council approve the recommended amendments to the Zoning By-law in this report, it is intended that the Planning Discussion Group will continue to meet regarding matters related to Site Plan Control and construction management."

The one local BIA was "A supporter of the project from day one" (Brissenden, 2016). The Bloor Annex BIA Chair Brian Burchell stated at the outset, "Based on our communication with [the developer], we're very optimistic that they'll propose an inventive, thoughtful destination space" (Simcoe, 2015). Overall, there was little input from the local BIA, which was in favour of the plan from the beginning, even before the development application was submitted. Kevin Ward concluded that BIAs place particular emphasis on gentrifying residential development by introducing art, bars and cafes, and by valuing physical heritage that promote the attractiveness of city spaces (Ward, 2010). The proposed development was always retail-focused, with a central public market with an abundance of small retain spaces, as well as "public art installations meant to preserve and enhance the area's eclectic commercial character" (Novakovic, 2016). By their very nature, BIAs are designed to brand areas in urban spaces and to promote destination marketing (Ruffin, 2010, p. 462). As such, there was alignment between the BIAs objectives and those of the developers.

City Planning modified the proposal to address the concerns of too much density, by which time, following the more intensive consultation exercise, the HVRA was in support of the project (Dexter & Orme, 2017). While the building height remained the same, the proposal was altered by increasing open space, public realm improvements, and heritage retention (Toronto and East York Community Council, 2017). The four neighbourhood associations acknowledged their success in modifying the plan:

> After three years and approximately 100 meetings with city staff, councillors, [the collection of neighbourhood associations], various communities and other stakeholders, Westbank's third proposal was unanimously approved by Toronto City Council on April 28, 2017. The changes made to the development plans are substantial. (Mirvish Village Task Group, n.d.)

They stated,

> In August 2014, before the formal project was unveiled, [the Palmerston Area Residents Association] did a survey in all four neighbourhoods. If you compare the wishes expressed by respondents to that survey, they bear a striking relationship to the document before you. Priorities set by the 276 respondents to the survey were greenspace, community cultural centre, family units, live/work, low income, small retail, market, arts, bikes ... community priorities that are imbedded in this re-zoning. (Dexter & Orme, 2017)

The changes included 13 fewer stories, reducing some of the density at the heart of the groups' concern (Dexter & Orme, 2017). They noted their role as "largely invisible," but one that "deepened our understanding of our communities. And that is city building at its best" (Dexter & Orme, 2017). The group intended to remain as a group in order to steward public engagement for the construction plan, site management, streetscape, and park design (Mirvish Village Task Group, n.d.), a role expressly endorsed by City Council (City Council, 2017a).

In summary, the City Planning Department, understanding the scope of the changes to be made in the neighbourhood, adopted a comprehensive consultation process that acknowledged the key roles of the resident and business associations in the area. However, the resident associations, rather than the BIAs, were the community partners in the Mirvish redevelopment. The sustainability part of the story was limited, superseded with resident concerns about density and traffic, although the potential for green space, heritage, and community culture played into the community visioning spearheaded by the neighbourhood association. The next vignette, which took place down the block a year later, tells a different story.

Bloor Street Bike Lane Pilot

In 2016, two downtown Councillors initiated a one-year pilot study for separate cycling paths (considered the gold standard) along a 2.4 kilometre section of Bloor Street between University and Shaw, a busy corridor for cycling, walking, and driving (Le Blanc, 2017). The Bloor Street corridor has long had extensive cycling and was previously identified as an important cycling corridor in the City of Toronto's Ten-Year Cycling Plan (Le Blanc, 2017). Cycling debates in Toronto have been divisive, with the city one of only a few that has removed cycling paths in the last decade. Advocacy groups such as Cycle Toronto have long been advocating more extensive, safer cycling infrastructure (Cycle Toronto, 2019). The city's downtown Councillors, who disproportionately favour separated bike paths, have had success by initiating pilot studies – an approach also used in New York, when it began installing cycling lanes.

The pilot study began in November 2016 and concluded with a request for permanence one year later (Aboelsaud, 2017). Two of the area's three BIAs (together, the "BIAs") were involved in the pilot and the debate for a permanent bike path from the start. The Bloor Annex Business Improvement Area and the Korea Town BIA commissioned a study in October 2015 on the local economic impact of bike lanes on Bloor Street (City of Toronto, 2017, p. 2). In October 2015, the Bloor Annex BIA and the Korea Town BIA commissioned a study on the local economic impact of bike lanes on a 2.4 kilometre stretch of Bloor Street to be carried out by an advocacy group, Toronto Centre for Active Transportation, in partnership with the University of Toronto and with support from the Metcalf Foundation (City of Toronto, 2017, p. 13). In May 2017, City staff released the Bloor Street West Bike Lane Pilot Project Evaluation, with a comprehensive review of findings (Aboelsaud, 2017). Staff cited a number of reasons for introducing permanent cycling lanes, including "reduced air pollution and greenhouse gas emissions, in keeping with Toronto's commitments in TransformTO" (City of Toronto,

2017b, p. 4). However, the main focus used when analysing the pilot was economic indicators and, specifically, the impact on local businesses.

The study looked at four economic indicators – customer counts, visit frequency, spending, and vacancy rates in the Bloor Annex and Korea Town business districts and, by all accounts, was enormously detailed. As staff noted, "The monitoring methodology employed for the Bloor Street pilot project has involved the most comprehensive performance evaluation undertaken for a cycling project in the City of Toronto" (City of Toronto, October 2017b, p. 5). It entailed looking at the local business economic impact along the applicable streets and nearby corridors, as well as the effects on parking and safety. The study included a door-to-door merchant survey and a pedestrian intercept survey (City of Toronto, October 2017b, p. 2). In addition, at the urging of local businesses, the city obtained customer spending analysis from Moneris Solutions Corporation, the company with the largest market share of point-of-sale payment processers in Canada. The findings were that, despite the removal of approximately 160 on-street parking spots and one traffic lane, business on Bloor Street continued to flourish during the pilot period (Dandy Horse, 2017) and that customer spending in the Bloor area increased (City of Toronto, October 2017b, p. 2). In addition, cycling increased, accidents were reduced, there was less conflict between motorists and cyclists, and visitors cycling to Bloor Street more than doubled from 7 per cent to 18 per cent.

The community engagement was widespread and extensive. City staff consulted the BIAs and three neighbourhood associations (Palmerston Area Residents Association, Annex Residents Association, and the Harbord Village Residents Association), held public consultation events, including one attended by over 330 participants, and fielded phone requests and held meetings with local businesses (City of Toronto, 2017b, pp. 16–17). By all accounts, resident and business organizations were enthusiastic. For example, the Mirvish Village BIA noted that the "BIA area is undergoing a huge transformation with the pending redevelopment of the Honest Ed's site and Markham Street" and that they "wish to emerge as a local community that includes the increased safety that bike lanes bring for all street users" (Mirvish Village BIA, 2017). Other organizations liked the idea of permanent lanes, too. The Bloor St. Culture Corridor, a collection of cultural organizations along Bloor Street such as the Royal Ontario Museum and the Royal Conservatory, saw the bike lanes as a way of providing "safer cycling transportation throughout our neighbourhood and to our cultural destinations and local businesses" and making member organizations more accessible to people from all areas of the city and with varying income levels (Bloor St Culture

Corridor, 2017). The Federation of North Toronto Residents Associations, representing dozens of neighbourhood associations in the northern part of the city, far beyond Bloor Street, also supported the initiative (Federation of North Toronto Residents Associations, 2017).

Amidst seemingly universal praise for the bike lanes was an underlying discord within the business community. An initial review of the opinions of about 140 local business representatives saw an even split in support and opposition (City of Toronto, 5 June 2017). There was generally less support amongst the businesses represented by the Korea Town BIA (Spurr, 2016). While these concerns were placated through modifications to the lanes and parking, a few months before the issue went to the Public Works and Infrastructure Committee for final decision-making and then City Council, a new business advocacy organization called the Annex Business Bike Alliance (ABBA) was created (IBikeTO, 2017). ABBA was not a formal BIA, a process that takes upwards of a year to create, but was instead an informal collection of businesses already represented by the local BIAs. ABBA cited the negative impact of the lanes on local sales and proposed changing the design of the lanes and introducing operating hours. The group stated that "the lanes were rammed into being by a group of ideologues," including the local Councillors and the Bloor Annex BIA, with a biased group leading the study. ABBA argued that they "came together in March [2017] after repeated efforts to work with our local BIA failed" (Annex Business Bike Alliance, 2017). According to one journalist, "The fix was in from the start and the survey results will not reflect in the slightest what many Bloor St. business owners are really experiencing" (Levy, 2017). ABBA's opposition called into question the degree to which BIAs were aligned on this project, how much consultation had taken place, whether the study was accurate, and the conflict of interest in having only some BIAs participate in survey design. In the end, following approval of the bike lanes amidst this discord, City Council ordered that ABBA, not the official BIAs, serve as the community to be consulted in the implementation phase (City Council, 2017b).

This case illustrated how local advocacy groups can go beyond participation to spearheading community projects in local sustainability projects. In this case, the BIA participated in the design of the study that would become crucial in receiving assent for permanent bike lanes. Although BIAs must be created through a complex, formal process in order to receive approval from City Council, informal business associations can assert authority and representation, too. Coupled with city staff's overall agnosticism on the formal structures of resident and business associations, this opens up significant questions about how these organizations can be accountable to their members – and to the public.

It also raises questions about the relationship between BIAs and sustainability projects, suggesting that "green" objectives can fade into the background where these bodies lead.

Stakeholder Group Overlap, Representation, and Accountability

This chapter raises more questions than it answers. Even so, I hope that it offers a glimpse into the ways in which local associations affect local sustainability initiatives. Several conclusions can be reached on sustainability initiatives and the overlap of local associations. First, the case studies reveal the extent to which the engagement process is *ad hoc*, as suggested by Reese et al (2001). In the Mirvish redevelopment, city staff created a stakeholder advisory group following the implementation of a consultation plan in order to be nimble, to be able to go back and forth between the parties involved – developer, residents, local businesses, and others. In the Bloor bike lane example, relying on BIAs and neighbourhood associations to represent their members ultimately backfired when ABBA came onto the scene, claiming to represent a different set of voices. ABBA questioned the objectivity of the local BIAs, given their leadership in the project. Interestingly, this example also showcases the extent to which the formality of BIAs may not elevate them to a privileged position when vis-à-vis city officials, in that City Council ultimately required that city staff consult with the ABBA on changes to the cycle track design, in advance of the creation of the permanent bike lanes, not the more formally created BIAs. ABBA's authority as a local business association representing its members, even though it was not a BIA, has echoes in Layard's view of "local" as asserting its own constructed reality, regardless of the formal instruments of law. In both vignettes, BIAs and neighbourhood associations were the chief stakeholder groups that the city used to facilitate public participation, even though these groups represent a particular, narrow set of interests.

Second, the cases illustrate that local advocacy groups were central to the development process, both as bodies engaged and active participants. Formal structures should carefully include the groups involved in local advocacy efforts. In the Mirvish redevelopment, one neighbourhood association played a central role at all stages of the process, from the announcement of the sale to the implementation of the plan. Although other groups, including BIAs, residents, and local businesses were involved in the city's stakeholder group, the voices of the neighbourhood associations were far more pronounced. The fact that a BIA was in the area and invited to participate in the stakeholder group does not mean that its degree of involvement was similar to that of the neighbourhood association. Likewise, the Bloor bike lane example shows the degree to which local BIAs went beyond acting as organizations to be consulted to partners in directing the project. Given this overlapping of roles, it calls into question their accountability to their members – are they crafters of public realm projects, respondents to city policy positions, or both? Researchers and practitioners should be conscious of the difference between resident and business associations responding to sustainable development initiatives, and their active participation in civic action. In addition, the city's identification of these groups as central to development demands greater attention to how they are accountable to their members, local residents, and broader stakeholders. Such considerations should form part of the design of urban engagement structures across Canada, including sustainability initiatives.

This chapter also sheds light on the many meanings and roles of sustainability in the planning process. In the Mirvish redevelopment, the concerns of the neighbourhood associations focused on balancing densification and an influx of new residents with the impact on development on existing residents. On the one hand, the extensive participation by resident associations in the planning process, the resulting reduction of a significant number of new rental spaces, and the emphasis on concerns such as traffic are consistent with literature on the reluctance of resident association support for significant new density. In this sense, sustainability implied a focus on current residents. On the other hand, neighbourhood associations were willing to endorse thousands of new residents in rental housing in return for amenities such as day care centres and park space, running counter to the notion that such bodies are reluctant to include lower-income residents and considerable increases in population. In this initiative, the local BIA supported the plan from the beginning, and densification was an attractive proposition with which to encourage business growth. Sustainability here meant support for urban density. The Bloor bike lane debate raised questions about the degree to which local economic factors dominate environmental reasons in relation to local infrastructure development, even where a project has sustainability at its core. Ultimately, in both cases, the local effects, including the quality of life for existing residents, the economy, and jobs, became the heart of the conversation, not greening the city. This raises questions about how local stakeholders dilute or advance the initial promise of sustainable initiatives by demanding modifications to plans that serve their members, but not the broader public.

This chapter has tried to show how local urban actors can affect governance in one particular urban context. In particular, it attempts to showcase the nuanced role that resident and business associations play in local decision-making. Zooming into specific spaces allows for a deeper understanding of how these bodies overlap with one another and impact debates.

Acknowledgments

I am grateful to Dayna Nadine Scott, Stepan Wood, and Sonia Lawrence, and especially Hoi Kong and Tanya Monforte, for their impeccable contributions to the analysis in this chapter. All errors and omissions are my own.

REFERENCES

Aboelsaud, Y. (2017, 11 October). City staff recommend to permanently keep Bloor Street bike lanes. Daily Hive. http://dailyhive.com/toronto/bloor-street-bike-lanes-final-report-2017

Agyeman J., & Evans, T. (2003). Toward just sustainability in urban communities: Building equity rights with sustainable solutions. *The Annals of the American Academy of Political and Social Science, 590*(1), 35–53. http://doi.org/10.1177/0002716203256565

Alarcon De Morris, A., & Leistner, P. (2009). From neighborhood association system to participatory democracy: Broadening and deepening public involvement in Portland, Oregon. *National Civic Review,* 47–55. https://doi.org/10.1002/ncr.252

Alexander, G., & Peñalver, E. (2009). Properties of community. *Theoretical Inquiries in Law, 10*(1), 127–160. https://doi.org/10.2202/1565-3404.1211

Annex Business Bike Alliance. (2017, 6 November). Summary: It's harder to operate a business on Bloor Street with THESE bike lines.

Berry, J. M., Portney, K. E., & Thomson, K. (1993). *The rebirth of urban democracy.* Brookings Institution Press.

Bloor St. Culture Corridor. (2017, 14 October). Correspondence to Public Works and Infrastructure Committee.

Briffault, R. (1990). Our localism: Part II – Localism and legal theory. *Columbia Law Review, 90*(2), 346–354. https://doi.org/10.2307/1122776

Brissenden, A. (2015, 16 February). Mirvish Village architect has social justice, grassroots background. The Gleaner Community Press.

Brissenden, A. (2016, 25 July). City hosts consultation on Westbank's Mirvish Village plan. The Gleaner Community Press.

Buckman, S. T. (2011). Upper middle class NIMBY in Phoenix: The community dynamics of the development process in the Arcadia neighborhood? *Journal of Community Practice, 19*(3), 308–325. https://doi.org/10.1080/10705422.2011.595306

Carmona, M., & de Magalhães, C. (2009). Local environmental quality: Establishing acceptable standards in England. *The Town Planning Review 80*(4/5), 517–548. http://doi.org/10.3828/tpr.2009.9

Chaskin, R. J., & Garg, S. (1997). The issue of governance in neighborhood-based initiatives. *Urban Affairs Review, 32*(5), 631–661. https://doi.org/10.1177%2F107808749703200502

Chaskin, R. J., & Greenberg, D. M. (2015). Between public and private action: Neighborhood organizations and local governance. *Nonprofit and Voluntary Sector Quarterly, 44*(2), 248–267. http://doi.org/10.1177/0899764013510407

Chemerinsky, E., & Kleiner, S. (2013). Federalism from the neighborhood up: Los Angeles's neighborhood councils, minority representation, and democratic legitimacy. *Yale Law & Policy Review, 32*(2), 569–581.

Chisholm, S. (2004). The growing role of citizen engagement in urban naturalization: The case of Canada. *Ekistics, 71*, 35–44.

City Council. (2017a, 26–28 April). Honest Ed's and Mirvish Village 571 to 597 Bloor Street West, 738 to 782 Bathurst Street, 26 to 38 Lennox Street, 581 to 603 and 588 to 612 Markham Street: Official plan amendment, zoning by-law amendment applications – Final report. City of Toronto.

City Council. (2017b, 7 November). Bloor Street West bike lane pilot project evaluation. City of Toronto.

City of Los Angeles. (2017). Neighbourhood councils empower LA.

City of New York. (2017). About community boards.

City of Toronto. (2016). Business improvement areas (BIAs): Board governance structure.

City of Toronto. (2017a, 5 June). Feedback survey #2. City of Toronto.

City of Toronto. (2017b, 3 October). PW24.9 Bloor Street West bike lane pilot project evaluation. Public Works and Infrastructure Committee.

City of Toronto Councillor or staff member #1. (2016, 5 July). Anonymous interview, Toronto – author conducted.

City of Toronto Councillor or staff member #2. (2016, 7 July). Anonymous interview, Toronto – author conducted.

City of Toronto Councillor or staff member #3. (2016, 18 July). Anonymous interview, Toronto – author conducted.

City of Toronto Councillor or staff member #4. (2016, 18 May). Anonymous interview, Toronto – author conducted.

Cycle Toronto. (2019). Position statements.

Dandy Horse. (2017, 11 October). Report shows Bloor business is flourishing with new bike lanes.

Darchen, S. (2013). The creative city and the redevelopment of the Toronto entertainment district: A BIA-led regeneration process. *International Planning Studies, 18*(2), 188–203. https://doi.org/10.1080/13563475.2013.774147

Dexter, S., & Orme, C. (2017, 4 April). Letter to the Toronto and East York Community Council from the Harbord Village Residents Association.

Director Community Planning. (2017, 17 March). Honest Ed's and Mirvish Village – 571 to 597 Bloor Street West, 738 to 782 Bathurst Street, 26 to 38

Lennox Street, 581 to 603 and 588 to 612 Markham Street – Official plan amendment, zoning by-law amendment applications – Final report. City of Toronto.

Fagotto, E., & Fung, A. (2006). Empowered participation in urban governance: The Minneapolis Neighborhood Revitalization Program. *International Journal of Urban and Regional Research, 30*(3), 638–655. http://doi.org/10.1111/j.1468-2427.2006.00685.x

Federation of North Toronto Residents Associations. (2017, 15 October). Correspondence to Public Works and Infrastructure Committee.

Ford, R. T. (1999). Law's territory (a history of jurisdiction. *Michigan Law Review, 97*, 843.

Frug, G. E. (1980). The city as a legal concept. *Harvard Law Review, 93*(6), 1057.

Fuchs, E. (2012). Governing the twenty-first century city. *Journal of International Affairs, 65*(2), 43. https://www.jstor.org/stable/24388217

Fung, A., & Wright, E. O. (2001). Deepening democracy: Innovations in empowered participatory governance. *Politics & Society, 29*(1), 5–41. http://doi.org/10.1177/0032329201029001002

Government of the United Kingdom. Localism Act, 2011, c. 20. http://www.legislation.gov.uk/ukpga/2011/20/contents/enacted

Habermas, J. (1962). *The structural transformation of the public sphere.* MIT Press.

Hoyt, L., & Gopal-Agge, D. (2007). The business improvement district model: A balanced review of contemporary debates. *Geography Compass, 1*(4), 946–958. https://doi.org/10.1111/j.1749-8198.2007.00041.x

IBikeTO. (2017, 2 October). Adhoc Annex group's questionable self-survey claims to prove "dramatic" downturn in sales, yet wants to keep bike lanes anyway.

Jacobs, J. (1961). *The death and life of great American cities.* Knopf Doubleday Publishing Group.

Layard, A. (2012). The Localism Act 2011: What is "local" and how do we legally construct it? *Environmental Law Review, 14*(2), 134–144. http://doi.org/10.1350/enlr.2012.14.2.152

Le Blanc, S. (2017, 13 November). City Council votes in favour of keeping Bloor Bike lanes. *The Varsity.* https://thevarsity.ca/2017/11/13/city-council-votes-in-favour-of-keeping-bloor-bike-lanes

Levin-Waldman, O. M. (2013). Income, civic participation and achieving greater democracy. *The Journal of Socio-Economics, 43*, 83–92. https://doi.org/10.1016/j.socec.2013.01.004

Levy, S. (2017, 30 September). Bike lanes hurt Bloor Street shops. *Toronto Sun.*

Lewis, N. M. (2010). Grappling with governance: The emergence of business improvement districts in a national capital. *Urban Affairs Review, 46*(2), 180–217. http://doi.org/10.1177/1078087410378844

Lippert, R., & Sleiman, M. (2012). Ambassadors, business improvement district governance and knowledge of the urban. *Urban Studies, 49*(1), 61–76. http://doi.org/10.1177/0042098010396235

Mallett, W. J. (1993). Private government formation in the DC Metropolitan Area. *Growth & Change, 24*(3), 385–415. https://doi.org/10.1111/j.1468-2257.1993.tb00132.x

McCann, E. (2011). Urban policy mobilities and global circuits of knowledge: Toward a research agenda. *Annals of the Association of American Geographers, 101*(1), 107–130. https://doi.org/10.1080/00045608.2010.520219

Meek, J. W. (2008). Adaptive intermediate structures and local sustainability advances. *Public Administration Quarterly, 32*(3), 415–432.

Merriam-Webster. http://www.m-w.com

Miller, S. R. (2013). Legal neighborhoods. *Harvard Environmental Law Review, 37,* 105–166.

Mirvish Village BIA. (2017, 13 October). Correspondence to Public Works and Infrastructure Committee.

Mirvish Village Task Group. (n.d.). Welcome to the Mirvish Village Task Group (MVTG).

Morçöl, G., Vasavada, T., & Kim, S. (2014). Business improvement districts in urban governance: A longitudinal case study. *Administration & Society, 46*(7), 796–824. https://doi.org/10.1177%2F0095399712473985

Morçöl, G., & Wolf, J. F. (2010). Understanding business improvement districts: A new governance framework. *Public Administration Review, 70*(6), 906–913. https://doi.org/10.1111/j.1540-6210.2010.02222.x

Nicholson, L. M., Schwirian, K. P., & Schwirian, P. M. (2010). Childhood lead poisoning laws in New York City: Environment, politics and social action. *Children, Youth and Environments, 20*(1), 178–199.

Novakovic, S. (2016, 14 June). Mirvish Village: Updated plan praised, criticized at consultation. *Urban Toronto.*

Portney, K. E., & Berry, J. M. (1997). Mobilizing minority communities: Social capital and participation in urban neighbourhoods. *American Behavioural Scientist, 40*(5), 632–644. http://doi.org/10.1177/0002764297040005009

Province of Ontario. Corporations Act, R.S.O. 1990, c. C.38.

Reese, L. A., Ohren, J. F., Brinkerhoff, J. M., Mitchell, J., & Washington, C. (2001). What's wrong and what should be done? Comments on the Case Study. *Public Performance & Management Review, 24*(4), 425.

Ruffin, F. A. (2010). Collaborative network management for urban revitalization: The business improvement district model. *Public Performance & Management Review, 33*(3), 459–487. https://doi.org/10.2753/PMR1530-9576330308

Seaton Village Residents' Association. (2014, 2 October). Update.

Simcoe, L. (2015, 2 March). "Initial" plans for Honest Ed's development set to be unveiled. *Metro*.

Spurr, B. (2016, 23 October). Korea Town business owners "concerned" about Bloor bike lanes' impact. *Toronto Star*. https://www.thestar.com/news/gta/2016/10/23/korea-town-business-owners-concerned-about-bloor-bike-lanes-impact.html

Thomson, K. (2001). *From neighborhood to nation: The democratic foundations of civil society*. Tufts University Press.

Topping, D. (2015). Toronto residents' associations & neighbourhood groups map.

Toronto and East York Community Council. (2015, 6 October). TE11.35 – Preliminary report – Honest Ed's and Mirvish Village – 571 to 597 Bloor Street West, 738 to 782 Bathurst Street, 26 to 38 Lennox Street, 581 to 603 and 588 to 612 Markham Street – Official plan amendment, zoning amendment applications. City of Toronto.

Ward, K. (2010). Entrepreneurial urbanism and business improvement districts in the State of Wisconsin: A cosmopolitan critique. *Annals of the Association of American Geographers, 100*(5), 1177–1196. http://doi.org/10.1080/00045608.2010.520211

Wolf, J. F. (2006). Urban governance and business improvement districts: The Washington, DC BIDs. *International Journal of Public Administration, 29*, 53–75. https://doi.org/10.1080/01900690500408981

7 Complementing Citizen Engagement with Innovative Forms of Professional Co-production: A Renewed Case for Transdisciplinary Charrettes

NIK LUKA, BRIA AIRD, AND NINA-MARIE LISTER

How might governance mechanisms be (re)conjugated with social movements to tackle the climate emergency and intertwined "wicked problems" of the twenty-first century, given the complex challenges of collective action where feedback loops are often abstract and/or long and slow by nature?[1] This chapter complements the other contributions by focusing on the need for professional specialists to develop methods for transcending disciplinary boundaries in practice. Our main premise is that architects, landscape architects, planners, and engineers are widely seen as key actors in shifting towards sustainability – at least in principle.[2] Many of the decisions taken by what Corner (2006) calls "the allied design professions" end up shaping the flows of energy, materials, and users (both human and otherwise), creating path dependencies for generations or even centuries to follow (Moe, 2019; Sorensen, 2011; Waldheim, 2016). Moreover, like the civil-society actors and the wide array of specialists who generate empirical evidence for decision-makers, professional practitioners in the instrumental-interventionist work of planning and design play key roles at the front lines of struggles for healthy, sustainable metropolitan regions by (re)producing landscapes, buildings, and infrastructure options that

---

[1] Usually attributed to the work of Rittel and Webber (1973), and as developed by Buchanan (1992), Daviter (2019), Termeer et al. (2015), and Westley and McGowan (2020), the term "wicked problem" is widely used in policy, planning, and design to describe complex situations for which responses are elusive, contingent on extraordinary collaborative efforts, and highly challenging to implement.

[2] Many observers question the real agency of professional actors in contexts driven hard by capital and politics – e.g., Bird and Luka (2010), Fainstein (1999), Milroy (2009), and Zukin (1988, 2010) – but we write here from a position of cautious optimism.

tend to get locked in for generations (De Block, 2016; Easterling, 2010, 2014; Sorensen, 2011).

A recurrent and perhaps generic response to our opening question, as repeated time and again in current debates, is that cross-disciplinary collaboration will ensure that work on "wicked problems" can be done well; normative treatises on sustainability thus assert that interdisciplinary and/or transdisciplinary collaboration is key to successful processes (Dale, 2001; Neuman, 2016; Ramadier, 2004). Yet the work of co-production *among* "experts" with professional credentials tends to be less well documented than the exciting and rather more "media-friendly" work of co-production linking specialists *with* diverse publics. In response, this chapter draws on a quarter-century of professional practice in which the authors have been involved in Quebec and Ontario, as well as elsewhere in North America and Europe.[3] Our reflections are combined with a review of secondary sources to argue that a long-established mode of production in many design professions – the *charrette* – can help actors in various fields to move forward on "wicked questions" of co-production involving many stakeholders from different disciplines, professional cultures, and fields of practice.[4] The arguments we present here have been honed on a major study of design-focused action-research methods where hands-on peer-to-peer experiential learning was done through intensive workshops exploring the nature(s) of co-producing transformative practical knowledge.[5] Our claims are furthermore based on analysis

---

3 We have been involved in over 100 major projects involving co-production with diverse publics, professional experts, elected officials, and other stakeholders in various combinations. Much of what we describe here is based on what is being done in Montreal (where the lead author is a key actor, regularly called upon to speak on and/or play a key role in joint decision-making on landscape planning, policy development, and urban design). Montreal is of particular interest because of its approach to participatory governance, exemplified by the Office de consultation de Montréal, which vets all major policy and project strategies with diverse publics in inclusive processes that refine draft proposals; see, e.g., Aubin and Bornstein (2012), Bherer and Breux (2012), and Van Neste and Martin (2018).

4 We use the term "co-production" to denote the joint, non-hierarchical development of policies, strategies, and projects among diverse groups of actors, whether horizontal (across disciplines and professions) or vertical (involving specialists and non-specialists); see Aubin and Bornstein (2012), Bornstein (2010), Horrigan (2014), Kong et al. (2017), Luka (2018), Neuman (2016), Toker (2012), and Vardouli (2016).

5 Entitled "Safe Passage: Towards an Integrated Planning Approach for Landscape Connectivity," this action-research project was awarded a three-year Partnership Development Grant by the Social Sciences and Humanities Research Council of Canada (SSHRC), with a project budget of $500,000 from SSHRC, the Woodcock Foundation, the Western Transportation Institute (a joint endeavour of Montana State University and the Departments of Transportation for the States of Montana and California), and the

of in-depth interviews with over 30 decision-leaders in transportation, ecology, landscape architecture, and engineering, which confirmed the need for transdisciplinary problem-solving on complex challenges that extend beyond the purview of any one profession, agency, or jurisdiction (see, e.g., Lister et al., 2015; Bell et al., 2020). At the crux of our argument in favour of the charrette – which we define in a moment – is the need to consider certain (and very numerous) moments in collective decision-making processes where citizen-experts *cannot* typically be present for practical and other reasons. In other words, attention must be drawn to how *collaboration among specialists* can more usefully *complement* the necessary and generative work of public participation in decision-making. Just as it is important to analyse how diverse "non-specialist" publics are (and should be) involved in design, policy, and project development (see, e.g., Luka, 2018), critical scholarship must also explore how specialists interact with one another on problems of collective cross-sectoral action.

It is imperative for us to explain before proceeding that we do not by any means eschew public participation. Indeed, much of our own previous work (Farina et al., 2014; Kong et al., 2017; Luka, 2018; Luka & Lister, 2000) has both asserted the importance of citizen engagement and explored how to operationalize the normative aims outlined in the early work of Jacobs (1961), Davidoff (1965), and Arnstein (1969).[6] Insights arising since the "participatory turn" of the 1960s include warnings about seeking to bring non-specialist publics into *all* phases of policy and design development. Not only are there practical and logistical reasons for public-sector actors to be mindful about *how* and *when* to impose a participatory moment on diverse publics (given deep issues identified in the literature such as public fatigue, cynicism, and the inefficacy of large-scale engagement (see Bherer, 2006, 2010, 2011; Fischer 1993, 2006), there is the simple fact that mobilization of

---

ARC Partnership. Its aim is to develop integrated planning and design approaches to implement wildlife-crossing infrastructure for sustainability and enhanced landscape connectivity in urban and urbanizing landscapes across Canada. Its premise is that such infrastructure requires an integrated approach that entails the conjugation of disciplinary expertise from ecology, planning, landscape architecture, and civil engineering (Lister et al., 2015). Empirical findings confirm that this sort of work requires cross-sectoral collaboration, since the challenge does not fall cleanly under the mandate of any one agency, nor is its impact limited to single jurisdictions.

6 Methods for public engagement were subsequently developed by countless others, as discussed in two key texts originally published in the late twentieth century: John Zeisel's *Inquiry by Design* (2006) and Henry Sanoff's *Community Participation Methods in Design and Planning* (2000). Newer cross-disciplinary syntheses of note include Toker (2012), Van Herzele (2004), Wates (2008, 2014), and Zukin et al. (2006).

diverse publics is difficult when typical citizens – especially if they are active in the labour force – can be overwhelmed with regular daily activities (see, e.g., Bherer & Breux, 2012; Blanchet-Cohen, 2014; Bornstein & Leetmaa, 2015). Space does not permit a review of how different activities can be combined at various "upstream" and "downstream" moments, but a lively debate continues in the scholarly literature on this topic.[7] Instead, we acknowledge that public participation has gained long-overdue prominence in planning and design over the last three decades, and as argued by many observers (Bornstein, 2010; Gamon, 1991; Girling et al., 2006; Hou & Kinoshita, 2007; McAvoy et al., 2004; Sutton & Kemp, 2006; Zeisel, 2006), it is important, valid, and an exciting part of design practice – but it is not the *only* kind of co-production that matters. Bringing together design professionals, other key decision-makers, and diverse publics is imperative in collective action, but we suggest that there are reasonable grounds for distinguishing between focused activities among specialists and other "moments" of co-production with broader participation. Complementarity is the charmed term here! How can highly inclusive activities of "crowdsourcing" be conjugated with "narrower" (curated) participation among specialists to achieve a balance between the conflicting demands of public engagement and efficiency?

This chapter aims to address a perennial question in the work of co-production, one arising from a review that surprisingly revealed a dearth of published empirical work on how professional experts work collaboratively across disciplines. How should co-production be curated in ways that enable different actors to engage in meaningful ways at different "moments" via particular modes of engagement, on terms that make good sense to them and make good use of their specific kinds of expertise and/or legitimacy as stakeholders, given the scarcity of resources (especially time and energy) and the urgency of resolving specific concerns? As the cliché goes, such interdisciplinary and/or transdisciplinary collaboration is more easily "said" than "done."[8] To this

---

7 Notable contributions include Agyeman (2013), Bornstein (2010), Forester (1998, 2009, 2018), Gustafson and Hertting (2017), Hilbrandt (2017), Van Herzele and van Woerkum (2008), and Van Neste and Martin (2018).
8 For instance, Ann Dale (2001) compellingly argued that collective action for sustainability entails transforming the silos, stovepipes, and solitudes engendered by disciplinary absolutism in theory and practice – as illustrated in the case-study chapters in this collection, especially in terms of the insights arising from grassroots initiatives that transcend the conventional nexus of the state, other institutional interlocutors, and diverse publics (who may have limited agency in decision-making processes). Getting beyond absolutist notions of disciplinary "purity" has also been identified by Alexis Shotwell (2016) as a fundamental challenge for the twenty-first century.

end, the introduction to this collection rightly asserts that where policy gridlock is an issue, there is a need for law and "concrete" processes of policy development to enable social movements to play more effective and meaningful roles in guiding change. We are thus especially interested in governance mechanisms that offer potential for reconciling what may seem to be incommensurable values, and for providing platforms for social movements that often face obstacles in being brought to bear for the public good. We nevertheless take a normative stance by asserting that in liberal democracies, processes of public engagement on matters of governance – which must, by definition, involve the widest diversity of stakeholders, irrespective of their formal "expertise," because they are members of the polities affected and/or who merit having voice and agency in the spirit of subsidiarity – must be matched by robust mechanisms for ensuring that specialists are empowered to interact with each other across the disciplinary and institutional systems that render "striated" (atomized, fragmented, exclusionary) situations "smooth" (inclusive), to use the conceptual vocabulary of Deleuze and Guattari (1980). Drawing on practical models as described by Bos et al. (2007) and as employed by Dale et al. (2013), we present the charrette as exemplar for optimizing the co-creation of knowledge and facilitating its transfer among researchers, state decision-makers, and design professionals, in ways that complement the all-important work of public participation writ large. We then argue that the time has come to redefine what a charrette is – and can be – in terms of the cross-disciplinary work of policy-making, planning, and design development across North America. The chapter culminates with a set of insights offered by charrette-focused modes of co-production vis-à-vis the general aims of this collection combined with an illustrative case from Montreal. In offering a critical reflection on what charrette-based models of co-production can offer beyond their usual stomping grounds of architecture, urban design, and landscape architecture, we suggest specific recommendations for the areas of praxis common to urban studies, planning, design, law, and public administration, where resilience and sustainability are important yet elusive goals.

What Is a Charrette?

Best described as a sort of "workshop" that is concentrated in time and space, a charrette seeks to infuse other modes of project and policy development with important bursts of energy and fresh insight. The charrette is a familiar method for architects, landscape architects, urban designers, and planners (see notably Beudin, 2015; Girling et al., 2006;

Roggema, 2014; Stevens, 1998) – acknowledged as a vital if endangered part of professional practice.[9] The term itself has peculiar double meanings, both of which concern the intensity of work done vis-à-vis the time allocated. To be "in charrette mode" means to be working furiously, pulling out all of the proverbial stops to produce a set of deliverables for a project. It characterizes the final 24 to 72 hours (often without sleep) before a deadline, as seen in many areas of professional endeavour where hard deadlines are inherent to practice.[10] It might seem odd to speak of this in anything other than pathological terms, but many practitioners come to be somehow addicted to the adrenalin-fuelled rush of working at full pitch to produce something important for a client or project leader. It is, of course, complemented by slow periods of contemplation, reflection, and rest (especially immediately after a deadline). This is not the notion of "charrette" to which we refer here. A second use of the term in design practice concerns a formal event intended to infuse other modes of design development with important bursts of energy and fresh insight, where a carefully curated group of specialists come together to tackle a specific question or problem that cannot be resolved through conventional work processes.[11] This latter understanding is our focus here.

The design charrette gained currency for design development in American design circles in the 1950s; the *Oxford English Dictionary (OED)* tells us that its first documented written use in English is from a 1959 article in the *Journal of Architectural Education,* but it was widespread long before that (see notably Beudin, 2015, and Stevens, 1998). In recent decades, it has come to be strongly associated with the "New Urbanism," an architect-led movement in the United States, to some extent monopolized by firms and non-profit organizations specializing in project management such as the National Charrette Institute (Condon, 1996, 2008; Condon et al., 1999; Evans-Cowley & Gough, 2009;

---

9 See Willis (2010) for a thoughtful discussion of the struggles designers now face to maintain creative practices in the increasingly capital-driven context of the Anglo-American world.
10 For instance, comparable modes of production can be seen in journalism (with a countdown to broadcast) or law (where practitioners work to set court dates).
11 This notion of the charrette is thus comparable to design competitions, albeit in less elaborate forms, which are similar to tendering in other sectors. Firms and/or consortia prepare submissions of ideas and schemes for adjudication by a jury of experts – usually with tight limitations on time – in the hopes of securing the commission for the project in question. See Chupin et al. (2015), Lehrer (2011), Lister (2012), and Waldheim (2016).

Kelbaugh, 2011; Lennertz, 2003; Walters, 2007).[12] Yet the popularity of design charrettes in the Anglo-American world since the 1970s has been matched by a broadening of meaning that we see as confounding the very alluring possibilities of this approach for transformative cross-disciplinary action. To support this claim, consider first the authoritative *OED* (the closest thing in the English-speaking world to the Académie française), which contains a draft addition from 2007 comprising three new definitions based on contemporary usage in North America (OED, 2018):

- "a period of intense (group) work, typically undertaken in order to meet a deadline,"
- "a collaborative workshop focusing on a particular problem or project," and
- "a public meeting or conference devoted to discussion of a proposed community-building project."

The last notion, linked with landscape planning and urban design, is quite important. It is not uncommon for community-engagement activities to be billed as charrettes, satisfying legislative requirements for due process on project development by engaging members of "the public" or "the community" (howsoever defined). Nevertheless, there are many moments where diverse publics cannot easily be integrated into detailed discussions, especially when due process requires a high level of technical knowledge. Moreover, in terms of what Agyeman (2013) calls *just sustainabilities*, there is something unsettling about purporting to engage a representative array of citizens in a process while only working with a handful of what are often the "usual suspects" of well-educated individuals from the dominant ethnocultural group in a given geographic context.[13]

We suggest that a semantic distinction be drawn anew between a participatory "workshop" involving a wide array of stakeholders – whether representative or not – and a "charrette" among specialists, thus acknowledging the focused notion that the term first seems to have denoted in the days of the École des Beaux-Arts. Such parsimony is seen in the work

---

12 Previously based in Portland (Oregon), a hotbed of "smart growth," the NCI is now housed within the School of Planning, Design, and Construction at Michigan State University; see http://www.canr.msu.edu/nci/ and Lennertz and Lutzenhiser (2006). On New Urbanism, see Grant (2006) and Moore (2013).
13 We refer here to the pitfalls of tokenism as discussed by Arnstein (1969); see also Fung (2003).

of Roggema (2014, p. 20), who stresses that a charrette need not involve the general public, but rather is marked by embracing "bottom-up" modes of co-production that "take local knowledge and perceptions into account in designing and decision making."[14] He also argues that a charrette is defined by six attributes (pp. 19–20):

1 Integrates intuitive, rational and emotional knowledge;
2 Is an inventive approach, includes idea-generating forces and results in envisioning futures;
3 Is set up in a creative atmosphere to allow many different stakeholders to collaborate;
4 Alternates between plenary discussions and small mixed design teams to provide a creative environment to think about the future in unlimited ways;
5 Creates an environment in which out-dated frameworks, often related to individual beliefs or "siloed" [sic] policies, can be overcome;
6 Makes use of maps and other visual tools to allow people to collaborate and integrate topographical, ecological as well as social and economic aspects.

To summarize: a charrette is an intensive, transdisciplinary mode of production among participants who are *deeply* involved in a given decision-making process; it complements other modes of co-production that are based on *broad* public involvement. Its basic "product" is design development. In the next section, we offer a reflection on how design charrettes can be reclaimed as effective and powerful components of professional practice that help to empower other modes of co-production involving diverse publics.

The Charrette as a Mode of Transdisciplinary Co-production

Many have explored the importance of competitions and charrettes for design practice (Chupin et al., 2015; Cross, 2007; Cuff, 1989; Hughes, 2017; Kelbaugh, 2011; Lehrer, 2011; Lister, 2012; Schluntz, 1982; Schröder, 1993; Shaw, 2002), and we do not wish to rehearse these arguments here. They seem useful for bringing intersectional knowledge

---

14 Without putting too fine a point on the matter, one might say that expecting diverse publics to participate in certain aspects of a process can unnecessarily render this transformative approach cumbersome while also discounting the agency of professionals in decision-making processes.

to bear on complex projects in architecture, planning, and landscape management, and we therefore take as a premise that intensive modes of co-production with a diversity of specialized participants have intrinsic value (in general terms) for the processes of design development. The nature of these benefits will vary from project to project, of course, but the "pressure-cooker" approach endures in practice. Exploring how charrettes – as time- and place-specific events for co-production among stakeholders – can enrich transdisciplinary work affords us the opportunity to think critically about how actors can participate directly in the formulation and implementation of initiatives for sustainability and resilience. The following paragraphs present a case for how charrettes (as focused activities among specialists) should be differentiated from instances of co-production involving diverse publics working in concert with professional "experts."

Various procedural characteristics are identified in the scant published work on charrette processes, whether theoretical or empirical. As a collaborative exercise, a charrette aims to unite key actors in unorthodox ways through a "burst" of activity in which "normal" ways of thinking and acting are questioned if not altogether set aside. It is usually an intensified, accelerated, and cathartic moment in which stakeholders are invited to step out of their "comfort zones" (i.e., their conventional silos of practice) – an event predicated on some sort of useful "shock doctrine." By this, we mean both praxis (in the classical sense) and *habitus*, as Pierre Bourdieu (1977) defined it (i.e., unselfconscious ways of doing that perpetuate themselves, whether or not they remain useful).[15] Architects, planners, engineers, developers, and other agents in design processes can easily fall into patterns of thought and action that are problematic – perhaps because they are based on "business-as-usual" ("best-practice") ways of doing things, perhaps because they are simply contradictory and counterproductive vis-à-vis the complex objectives in a given project, or perhaps because they represent a path of least resistance in institutional terms.[16] When done well, the charrette represents an opportunity to break out of these path dependencies by forcing stakeholders out of their familiar silos of practice – to become transdisciplinary in their working method.[17] It thus challenges the "normal"

---

15 On praxis, see Amador et al. (2015), Anderson (2014), Fainstein (1999), Milroy (2009), and Moe (2019).
16 An important critique of "best-practice" approaches is found in Moore (2013).
17 Designed to stimulate collaboration, a transdisciplinary approach can also narrow the gap between research and decision-making (Brown, 2010; Ramadier, 2004).

(discipline-specific) way of doing things in light of the problems of interdisciplinarity as articulated by Ramadier (2004), i.e., operating under the Enlightenment belief in the fundamental unity of knowledge and thus seeking to simplify and generalize knowledge among disciplines to create more robust knowledge about objects. By revealing the tensions and conflict at disciplinary boundaries, the charrette is an activity at the junctures of professional/disciplinary specializations, akin to what Cindi Katz (1996) called "minor theory" (operating at the margins of established canons and stories about how things ought to be done), in which coherence, rather than unity, is a primary goal (Ramadier, 2004).[18] Charrette work is therefore congruent with Rapoport's (1995) definition of design as problem-solving, for it aims to work through problems of process in the design project itself. It does not presume, however, that design is solely the work of "solving" problems; when done well, a charrette serves to reveal context-specific and situational opportunities that may not have been apparent to specialists working in "business-as-usual" spaces of practice.

Prominent among the benefits associated with charrette processes, if the mix of participants is well curated, is the possibility of "breakthrough" – overcoming procedural and epistemological boondoggles. It is also clear from empirical work, however, that mutual trust, empathy, and familiarity are all preconditions for this to occur (Forester, 1998, 2006, 2009; Gordon & Baldwin-Philippi, 2014; Hou & Kinoshita, 2007; Laurian, 2009; Swain & Tait, 2007). This entails developing appropriate levels of knowledge, trust, and openness to different ways of thinking. Most accounts of charrettes or workshops warn practitioners about the importance of being well prepared, and many specify that ill-preparedness can include trying to involve too many different sorts of expertise, howsoever recognized (Condon, 1996, 2008; Condon et al., 1999; Lennertz & Lutzenhiser, 2006; Sanoff, 2000; Sutton & Kemp, 2002, 2006; Wates, 2008, 2014).

In sum, it seems useful to specify that an event based directly on the principles of participatory democracy, distributed decision-making, and/or critical service learning warrants being differentiated from the work of a "charrette" (see Bherer et al., 2015; Horrigan, 2014; Mitchell,

---

18 Katz (1996) drew on poststructuralism, assemblage theory, and the work of Kafka on "minor literature" to decompose the "major" or dominant narratives of both theory and practice, challenging the notion of "mastery" that defines discipline-driven approaches to organizing anything from building transportation infrastructure to developing after-school programs. See also Shotwell (2016).

2008; Rosol, 2010; Sutton & Kemp, 2002, 2006; Toker, 2012; Wates, 2008, 2014). Beyond the generic notion of charrette as a moment of intense design development (working in *charrette mode*), we argue that two collaborative types of activity matter:

- a *transdisciplinary charrette* – understood as a collaborative, cathartic exercise of great intensity, focus, and decision-making defined in time and space as an event intended to move a project forward in some significant way, involving discrete kinds of specialist participants; and
- a *community workshop* – a decidedly public exercise in design development uniting "expert" and "lay" populations in carefully organized collaborative events that have even greater intensity and choreography than the charrette as a concerted event involving professionals and designated decision-makers, and that by definition are major acts of participatory democracy, comparable in terms of resource allocation and potential impact to the work leading up to referenda and elections.

We call for this differentiation based on the work of Forester (2018), Toker (2012), and Wates (2008, 2014) in acknowledgment of how cumbersome it can be to involve all manner of stakeholders – each a specialist in certain respects, but also a non-specialist in many others – in each part of co-production for collective action. The two types of activity represent different moments in design development, and they complement one another; what is more, they need not take place in a specific sequence. Both must be carefully situated in decision-making processes. For instance, a community workshop with broad (open) participation entails a major investment of resources, and as with all forms of participatory democracy, it should be undertaken only when circumstances can provide some assurance that it will be robust, accessible, meaningful for its participants, and effective (Bornstein, 2010; Forester, 1998, 2006, 2009, 2018; Fung, 2003; Laurian & Shaw, 2009; Swyngedouw & Moulaert, 2010). It is an ideal event to position both "upstream" and "downstream" of a specialist charrette, for public input early in a discovery process is key, and ideas developed during a charrette – as an "in-camera" exercise – should then be vetted with diverse publics afterward. Indeed, bringing a wide array of different actors together in one room for a charrette with the assumption that "we can work it out" – especially when this includes diverse publics and professionals with widely varying forms of what Agyeman and Erickson (2012) refer to as "cultural competency" – seems idealistic.

The rest of this chapter focuses on the transdisciplinary charrette as a mode of production in which key stakeholders in specialist roles come together to deliberate on potential courses of action for subsequent verification (e.g., through broad-based "crowdsourcing" exercises). Our call to "redefine the charrette" is not at odds with broad-based public participation in design development. In fact, it allows for complementarity and helps with the *débroussaillage* of identifying specific methods or events that are needed in the work of moving towards greater sustainability and resilience.

Key Advantages of Transdisciplinary Charrettes

The literature is surprisingly sparse in providing empirical evidence of how the charrette process matters, but we have identified a series of key points based on our engagement with participatory processes in combination with a review of normative treatises and critical (empirical) reflections. As one might expect, a prominent and unanimous claim concerns *the value of the "burst" or "compression" effect*, in shocking participants out of their business-as-usual ways of thinking, and the catharsis that stems from this intensity. This is not discussed in abstract terms, but rather with reference to (1) grounding procedural discussions in specific cases, (2) producing discrete deliverables, (3) generative "breakthroughs" of creative problem-solving that yield a useful vision or strategy, and (4) the crafting of a project-specific methodology (Condon, 1996, 2008; Condon et al., 1999; Gamon, 1991; Kelbaugh, 2011; Lennertz, 2003; Miller, 1982; Roggema, 2014; Schön, 1993; Sutton & Kemp, 2006; Walters, 2007; Wates, 2008, 2014; Willis, 2010). Advocates insist that by pooling energy and effort in short time frames (preventing redundancy and duplication of work), charrettes also offer useful and stimulating – even fun – alternatives to time-consuming meetings or conferences, especially as ideas are often "deposited" in these contexts only for passive consumption. The "compression" of space and time is seen as beneficial not only by accelerating change, but also because it can *minimize the need for re-working strategies due to misunderstandings among stakeholders* (Lennertz & Lutzenhiser, 2006; Osterwalder & Pigneur, 2010). In effect, the charrette "moment" demands that its participants provide clearer explanations, distilling jargon-laden discipline-specific narratives into accessible stories, building consensus by developing common vocabularies and frames of reference – ultimately yielding a shared vision of how to proceed (Lennertz & Lutzenhiser, 2006; Miller, 1982; Osterwalder & Pigneur, 2010; Roggema, 2014). This has the advantage of helping to prevent stalling tactics and to induce what Weinstock (2013) calls "principled moral compromise" – which is important where the development of "optimal" solutions is less important than satisficing.[19]

The intense, time-constrained, and cross-disciplinary conditions of a charrette create *tighter feedback loops in the design process* (Lennertz 2003; Lennertz & Lutzenhiser, 2006; Nassauer & Opdam, 2008; Roggema, 2014; Schluntz, 1982). These help to strengthen the quality of ideas and claims by forcing protagonists to render their rationales plain and explicit among peers; continuous exchange also produces formal knowledge that is both practicable and relevant to broader stakeholders (Nassauer & Opdam, 2008). Ideally, one sees simultaneous brainstorming, negotiation, and debate with feedback from peers throughout the process (Kelbaugh, 2011; Lennertz, 2003; Miller, 1982; Roggema, 2014). Indeed, the charrette is ideally a space *where disciplinary boundaries are transgressed and bridged through the project in hand*. Rather than merely providing a space of fleeting "encounter" among different specialists, the charrette process is one where transdisciplinary work develops new epistemic communities through deliberative encounters (Klein, 2008; Lennertz & Lutzenhiser, 2006; Roggema, 2014). It also reveals the iterative nature of design work, which shifts between concepts and empirical knowledge, alternately broadening and narrowing the problem space, engendering cycles of creating, testing, and recreating prototypes or scenarios (Cross, 2007; Osterwalder & Pigneur, 2010; Razzouk & Shute, 2012). In application, then, charrettes invite diverse participants to manipulate and change the problem frame to find satisfactory "solutions" through "what-if" explorations, thus allowing serendipity, leaps of faith, and simple intuition into the design-development process (Cross, 2007; Osterwalder & Pigneur, 2010; Roggema, 2014; Sutton & Kemp, 2006).[20]

Charrettes engender *mutual understanding and the deepening of participant capacity for meaningful deliberation* – which resonate with procedural observations found in the literature on transdisciplinary project development (Gordon & Baldwin-Philippi, 2014; Klein, 2008; Polk & Knutsson, 2008; Roggema, 2014). It is argued that this is possible because rigid and/or hierarchical relationships are rendered more "horizontal" though direct peer-to-peer dialogue (Lennertz & Lutzenhiser, 2006; Roggema, 2014). When contrasted, for instance, with standard institutional

---

19 Rather than seeking to optimize outcomes, design practitioners often seek to "satisfice" by mining competing alternatives for new possibilities, thus avoiding the "necessity of choice" of "either/or" responses (Razzouk & Shute, 2012).
20 On framing and reframing as an exercise, see Schön (1993), Schön and Rein (1994), and Van Neste and Martin (2018).

procedures in which certain protagonists receive reports or mandates – from which they prepare response memos, which are forwarded to other protagonists, perhaps with a few formal meetings – a charrette forcibly creates spaces of direct engagement among specialists. This helps to increase mutual awareness and learning by educating participants across the normal silos of practice and expertise, familiarizing different stakeholders with what their peers do; it can therefore break stereotypes down and overcome the parochialism that disciplines can nurture (Margerum, 2002; Roggema, 2014; Schluntz, 1982). Another advantage is that charrettes can thus *render it unnecessary for specialists to base their actions on an anticipatory (speculative) understanding of what their peers in another "silo" might do* once they "receive" the dossier in conventional linear and/or hierarchical ways of developing a given project.

The *project-specific nature of charrette work* is of great importance. Perhaps we write from a self-interested perspective (since architecture, landscape architecture, urban design, and planning are often project-focused, as opposed to other fields where generalities and deductive reasoning are more prominent), but the charrette almost inevitably demands that its *participants ground broad principles and procedural norms in site-specific cases*. This brings specialist knowledge to bear on specific situations, "ground-truthing" in ways that help to determine how and where their expertise is salient (Osterwalder & Pigneur, 2010) and – more often than not – requiring that it be expressed in ways that make sense to the co-producers of that knowledge (Opdam et al., 2013; Russell et al., 2008). Grounding, prototyping, and visualizing different scenarios are all inherent to a robust charrette process, as asserted by Miller (1982, p. 29):

> The charrette, as a competition or design workshop, adds a spatial dimension to that process – a vehicle to surface a gamut of wild hunches, an impromptu explosion of ideas, a site-specific resource, a set of documents offered for comparison, a graphic representation on paper to add to the compendium of verbal and written analyses.

Questions that can be explored through a charrette process generally operate through a constructivist epistemology (recognizing the situatedness of the research collaborators vis-à-vis the cultural context within which the knowledge is being produced), consistent with the claim made by Thompson et al. (2017) that transdisciplinary knowledge must be situated in specific contexts if it is to be relevant.

The contextualized, constructivist work of a charrette matters in another important way. Roggema (2014) observes that a charrette requires that its participants both draw upon and develop several kinds of knowledge (emotional, rational, and intuitive). Evidence suggests that charrettes can thus bring about both *outward and inward transformations among participants by enabling them to engage in moments of self-scrutiny and reflexivity* (Gordon & Baldwin-Philippi, 2014; Schluntz, 1982). This helps to *build confidence and respect among specialists from different domains* who might not otherwise have interpersonal bonds of trust or loyalty. In effect, a charrette has procedural value not solely through empowerment – that is, by generating agency and capacity among its participants – but by enabling them to demonstrate to one another what "good faith" means among different specialists. Mutual trust, confidence, and fealty cannot be taken for granted, especially in large undertakings such as state-led projects, where different factions or constituencies inevitably have divergent agendas. Rather, it must be generated and actively maintained, and this is one of the most compelling reasons that charrette methods are cathartic for multi-stakeholder, cross-sectoral projects (Lennertz & Lutzenhiser, 2006). Indeed, a charrette space, when appropriately constructed, provides its participants with a safe space to think openly and to speculate through what Osterwalder and Pigneur (2010) call "what-if" brainstorming. Its short time frames and multiple disciplinary perspectives engender a de-facto emphasis on *salient* knowledge, which is understood by advocates of integrative, design-based research to mean *knowledge that is meaningful, rigorous, and applicable for all parties*, especially those who have not conventionally participated in structured decision-making procedures beyond their own (disciplinary) comfort zone, including stakeholders who work at the proverbial coalface (Nassauer & Opdam, 2008).

Rounding out our inventory of how charrettes matter are two points concerning the receptivity and political traction of a given initiative. Charrettes can be of strategic use for external relations, with symbolic advantages in PR huzzah. Convening a charrette among professionals serves as *a gesture to elected officials and diverse publics that the specialists are stepping out of their "business-as-usual" modes of operation to figure out a new strategy* vis-à-vis the broader goals of sustainability, resilience, and efficacy. Miller (1982, p. 29), for instance, commented on how the "potential for public awareness and interaction with those involved in decision-making is of enormous benefit to a city." The message to convey outwardly is that we are looking to do something transformative. If nothing else, this can help to raise general awareness of the extraordinary amount of work needed on complex projects done in the public interest.

Finally, the intense and cathartic "pressure cooker" of the charrette can be especially important in high-stakes projects and contentious situations because of how they *help to defuse smouldering tensions among different sorts of specialists* (Lennertz & Lutzenhiser, 2006). Whether because of stereotypes that arise (couched in terms of "us" versus "them") or rivalries of other kinds, multi-stakeholder initiatives are marked by differences of perspective. Given the context-dependence of transdisciplinary work, its orientation to real-world problems, and the imperative of including collaborators with diverse disciplinary frameworks, tensions are inevitable. Mobjörk (2010) notes that conventional approaches to problem-solving based on discipline-specific (instrumental) modes of rationality often fail to address immediate, context-specific, "real-world" challenges. The experiential compression of a charrette can *help reveal to participants how transdisciplinary work depends on value-rationality* (Polk & Knutsson, 2008); it also helps to demonstrate that successful resolutions to wicked problems cannot depend solely on instrumental or "objective" forms of rationality.[21]

Procedural Tips on Making Transdisciplinary Charrettes Work

Our case in favour of charrettes is set against an array of contemporary issues including climate change, political unrest (expressed in many contexts by the rise of new forms of populism), and economic volatility. Burch et al. (2014) have demonstrated that there are urgent needs to improve policy coherence among different agencies and levels of government to address complex socio-ecological problems. They require innovative ways of generating transformative knowledge for institutional change (Dale, 2001; Irwin, 2015; Newell et al., 2015). Constructive responses are furthermore needed, given the erosion of popular trust in centralized governance structures, expert knowledge, and objectivist scientific authority (Larsen, 2015; Mobjörk, 2010; Polk & Knutsson, 2008; Ramadier, 2004; Russell et al., 2008; Wickson et al., 2006). This is especially true as decisions are increasingly seen as legitimate only if made on the basis of meaningful interaction of concerned stakeholders across scales and levels of governance in ways that integrate knowledge, policy, and action for sustainability (Dale, 2001; Farina et al., 2014; Kong et al., 2017).

In response, we have argued here that broad (public) participation should be complemented by intense moments of focused yet transformative brainstorming among a select group of actors, whereby

---

21 Fry (2001) points out that transdisciplinary approaches engage directly with normative values and that rationalities and methods of a given project often respond to the values of project initiators and participants, rather than to ideals of objective "truth."

interdisciplinary work can be meaningfully superseded by *transdisciplinary* approaches. While focused on applications, transdisciplinarity also generates useful knowledge that extends beyond the specific case contexts in which it arises (Krishna, 2001; Russell et al., 2008; Wickson et al., 2006). We have suggested that the charrette approach is especially pertinent to transdisciplinary work on complex, multi-stakeholder projects that are intended to build institutional capacity and resilience in all its useful forms. We now propose a series of performance principles for how to make transdisciplinary charrettes work well. Scholarship on this empirical question is scant, so many of these observations come directly from our first-hand experience working on large, complex projects.

- *Offer participants a safe space.* Ensure that the space/process is comfortable for participants who might keep their cards close to their chest in conventional linear processes of project development. Remind participants that we are exploring, that we are not committing to specific pathways during the charrette, and that the space it represents is one of internal speculation through abductive reasoning and leaps of faith to be explored in detail later (i.e., allowing for "failing safely") as opposed to work done for (and before) "external" parties (e.g., elected officials or the general public). It is also important to remember that the charrette space can be intimidating to non-designers (Sutton & Kemp, 2006), and to accommodate their concerns or anxieties in this respect.
- *Aim to generate new forms of dialogue; strive for spontaneity.* Institutional pathways bedevil governance processes, but the charrette can force a friendly confrontation of divergent "norms" among contrasting views of what needs to be done. Time compression can work wonders for what Forester (1998, 2018) described as the practical politics of muddling through. A useful strategy to foster new forms of dialogue is to have the charrette take place in situ, or at least near a site that embodies the questions and opportunities under examination (Miller, 1982): this gives a realism to the work and allows for quick ground-truthing. Parallels can be drawn here with the growing literature on "walking methods" (see, e.g., Luka & Lilley, 2018; Macpherson, 2016).
- *Produce images, diagrams, and clear statements of action.* Deliverables matter in charrettes. The time limitation of a charrette requires that professionals get something down on paper before the deadline. This urgency helps to prevent protagonists from deploying stall tactics (by saying things like "We will have to look into that" – which is often a polite way of saying "We will not bother looking into that").

The deliverables should privilege images and diagrams, not texts, as expressions of what is often called "visual thinking" (Corner, 1998; Frederick, 2007; Osterwalder & Pigneur, 2010; Tufte, 2006). It is also important to find ways to oscillate between details (e.g., of a procedural nature) and broader concerns. A useful tactic is to get participants to think through different scales – ranging from the long-term "big picture" of comprehensive planning (often privileged by public-sector actors) to the immediate time frames and site-specific conditions of specific projects (often privileged by private-sector actors). A charrette can thus allow for simultaneous brainstorming and negotiation (at least by proxy).

- *Practise humility*. Remember to respect the professional's awareness of the locus of control! A charrette space can provide "social permission" to relinquish power vis-à-vis key aspects of process; humility also means demonstrating good faith by shedding the language of guidance, education, and pontification that professionals often use with peers versus that of openness to risk (Miller, 1982, p. 29). One might borrow from the approach of "empathy mapping" proposed by Osterwalder and Pigneur (2010, p. 131) to prepare an inventory of the "pain" and "gain" experienced by participants as they see, hear, think, feel, and do certain things with their colleagues. Awareness of non-verbal communication is also of particular importance (Kong et al., 2017).

- *Strive for true co-production*. Let it be understood that this is a work of mutual authorship defying the (linear) sequence of policy and project development, flattening out hierarchies of power/status. Gamon (1991), Lennertz and Luitzenhiser (2006), and Miller (1982) all stress that the charrette should serve as an equalization process in which different actors with diverse (and often divergent) agendas come together for cross-functional discussion. A similar principle is stressed by Osterwalder and Pigneur (2010) in their compendium of strategies for generating innovative business models. For the public sector, the advantages of a collaborative exercise include a significant infusion of professional third-party design expertise. By bringing "fresh eyes" to a site-specific project in a structured and deliberate manner, the allocation of scarce resources can be rendered effective by affording a deliberative process of on-the-spot peer review (of the preoccupations, intentions, policies, and possibilities afforded by the case). With its emphasis on producing several options through the development of design scenarios, a charrette thus mirrors the design competition (Lehrer, 2011; Lister, 2012; Schluntz, 1982) – but arguably in a much more useful way.

- *Situate the charrette carefully in the development of policies and projects.* Almost all accounts in the literature conclude that the likelihood of a charrette producing useful results is directly related to the care taken in preparation and the degree of involvement of charrette participants before the event itself. It is a "midstream" moment in a process, and it must therefore be very carefully situated in time. Applying this principle also entails striving for some sort of disciplinary "balance" (Sutton & Kemp, 2006, p. 136) and recognizing that non-designers may feel like the proverbial fish out of water. A charrette will be effective only when the participants are meaningfully enabled to engage with core issues or questions. This is one way in which a charrette works best when done among specialists rather than with diverse publics. Indeed, the assumption of "the more the merrier" is facile.
- *Remember that the charrette is about learning; build reflexivity into the process.* A charrette is space of semi-suspended disbelief so that learning can occur across disciplinary silos. Universities and training institutes can play a key role in preparing, delivering, and assessing the effects of a transdisciplinary charrette (Forsyth, 2006; Sutton & Kemp, 2006); at the very least, a post-event reflection exercise helps participants to grapple with questions about how they were challenged to think in new ways, to express core assumptions with others from different backgrounds, and so on.[22]

It is important to note that the charrette should not be framed by *pensée magique*. In other words, it is not a magic bullet. It requires a significant investment of time and energy, and it must be carefully curated. One must therefore beware of pseudo-charrettes that are really just publicity stunts – hollow gestures of due diligence intended to generate confidence among the public and/or senior decision-makers or project auditors. When done well, however, the friendly juxtaposition of stakeholders in a transdisciplinary charrette can bring about important transformations in projects and processes, generating a cohesive, broadly supported version of what might otherwise have been seen as viable and useful only by certain parties. The spontaneity and informality of the charrette "moment" can work wonders in creating a richer space for dialogue, enabling tighter feedback loops than would normally be possible.

---

22 Miller (1982, p. 29) argued that universities and/or other institutions can sponsor charrettes, thus offering a considerable gift to the city.

Conclusion: An Illustrative Case

We wind up this reflection with a concrete illustration of how transdisciplinary charrettes have been inserted into a planning process that includes broad public participation but that complements "crowdsourcing" with expert workshops in ways that deliver the benefits we have described above. The Green, Active, Healthy Neighbourhood (GAHN) project was led by the Montreal Urban Ecology Centre (MUEC), in collaboration with other civil-society groups and local government, and it was funded by several granting agencies (see Blanchet-Cohen, 2014). The GAHN project aimed to redefine neighbourhoods in ways that they support walking, cycling, and transit use (Luka & Engle, 2015). It thus engaged the kind of complex or "wicked" problem that we have suggested above as demanding transdisciplinary charrettes, enabling participants to meaningfully engage with the core issues at stake in a specific project. It also had two features that situate it well in the context of this collection's themes. First, as documented by Blanchet-Cohen (2014), it involved significant citizen participation; indeed, the MUEC has a distinguished place in the history of community-based social movements in Montreal.[23] As the lead institution on the GAHN project, the MUEC ensured that it was rooted firmly in Montreal's unique ecosystem of citizen engagement. The second feature of the GAHN project resonant with this volume's themes was, of course, its focus on operationalizing ideas of urban sustainability, consistent with the MUEC's long-standing mission of advocating for "green, inclusive, just, democratic, and inclusive cities" (Luka & Engle, 2015).

A detailed account of the multifaceted approach developed in the GAHN project cannot be offered here, but a few highlights bear mentioning. Central to the process were strategies for public outreach, participatory planning, and community-based design activities linking local citizens with other stakeholders to generate progressive plans and policies (see Luka & Engle, 2015, for further detail). Four pilot "local GAHN plans" were undertaken in different Montreal neighbourhoods, each of which was selected for its potential to demonstrate institutional reform in various contexts, entailing formal agreements between local civil-society organizations and state institutions. Yet the community-based design workshops were complemented by transdisciplinary charrettes among

---

23 Among the MUEC's many contributions to Montreal's thriving civil-society sector was its role in helping to maintain cooperative housing central neighbourhoods in the face of pressures from real-estate speculators (Luka & Engle, 2015).

different specialists. Both types of activities generated extensive empirical evidence vis-à-vis core issues and opportunities in each neighbourhood; explorations of when to position both types of events (upstream and downstream) yielded important results through "crowdsourcing" and experts working in a safe, depoliticized space away from the performativity of public meetings, enabling discussions among specialists that propelled ideas and propositions that emerged from the broad-based participatory initiatives.

An external analysis of the GAHN project's early phases suggested that "professionals in private- and public-sector roles ... [were] exploring ways to enfold the project's principles and values into their ways of doing things" (Luka & Engle, 2015). Indeed, the transdisciplinary charrettes with professional actors from various sectors – notably the municipal government in addition to practitioners of planning and design who advocate for progressive approaches – enabled the benefits of reflexive learning and institutional change that we have discussed in this chapter and that have been identified in the afterword, given its importance as highlighted in other contributions to this collection. The professionals were able to conduct quick but pithy "on-the-spot" reviews of the preoccupations, policies, and possibilities revealed by the GAHN project's mandate and by public outreach. This kind of transdisciplinary deliberation was undertaken in a spirit of humility, in terms that do not privilege any single discipline's frameworks or procedural norms, and in ways that reflect a "practical politics of muddling through" (Forester, 1998, 2018), as we have argued above. Finally, it is worth nothing that preliminary work was also done to take advantage of the MUEC's deep connections with the academy, including several of Montreal's major universities. Ad-hoc exercises were developed so that participants could participate in reflection events, hosted and facilitated by the MUEC's academic collaborators; these were highly fruitful. This point resonates with several of the case studies discussed in this collection. Following completion of the GAHN project, new work done through the collaborative continent-wide project mentioned in the introduction (see note 5) has yielded additional insights into how critical reflections after the event can be incorporated as a matter of course into co-production processes to ensure that reflexive learning can continue long after the "burst" or "compression" effects of a charrette have subsided. As it happens, an extension project in Alberta is intended to deepen this action-research perspective, but the global COVID-19 pandemic required delaying this phase of the research until late 2022. We end, therefore, with a promise of news to follow in the coming years!

Acknowledgments

Thanks are extended to many people who have contributed to the development of material presented here: Marie-Hélène Armand, Lisa Bornstein, Marta Brocki, Renée Callahan, Stuart Candy, David Covo, Salmaan Craig, Ann Dale, Anurag Dhir, Rosetta S. Elkin, Jayne Engle, Gorka Espiau, Cynthia Farina, Audray Fontaine, Véronique Gasser Sikora, Isabelle Gaudette, Pierre-Étienne Gendron, Stéphan Gervais, Jeremy Guth, Nils Hertting, Michael Jemtrud, Hoi L. Kong, Mary Elizabeth Luka, Robert Mellin, Kiel Moe, Christine Mondor, Tanya Monforte, Raquel Peñalosa, Amy R. Poteete, Margaux Ruellan, Jessica Ruglis, Renée Sieber, Darren Veres, and Daniel Weinstock. This work was supported in part by grants from the Fonds de recherche du Québec (Société et Culture), the Social Sciences and Humanities Research Council of Canada, and the Enhancing Learning and Teaching in Engineering initiative at McGill University. We also wish to recognize excellent RA support provided by Dominique Boulet, Jaimie Cudmore, Alanna Felt, Sumeet Kulkarni, and Daniel Schwirtz. Finally, thanks are extended to the anonymous reviewers whose constructive comments ensured that this chapter offers a useful contribution.

REFERENCES

Agyeman, J. (2013). *Introducing just sustainabilities: Policy, planning, and practice.* Zed Books/Palgrave Macmillan.

Agyeman, J., & Erickson, J. S. (2012). Culture, recognition and the negotiation of difference: Some thoughts on cultural competency in planning education. *Journal of Planning Education and Research, 32*(3), 358–366. http://doi.org/10.1177/0739456X12441213

Amador, F., Martinho, A. P., Bacelar-Nicolau, P., Caeiro, S., & Oliveira, C. P. (2015). Education for sustainable development in higher education: Evaluating coherence between theory and *praxis. Assessment and Evaluation in Higher Education, 40*(6), 867–882. http://doi.org/10.1080/02602938.2015.1054783

Anderson, N. M. (2014). Public interest design as praxis. *Journal of Architectural Education, 68*(1), 16–27. http://doi.org/10.1080/10464883.2014.864896

Arnstein, S. R. (1969). A ladder of citizen participation. *Journal of the American Institute of Planners, 35*(4), 216–224. https://doi.org/10.1080/01944366908977225

Aubin, R., & Bornstein, L. (2012). Montréal's municipal guidelines for participation and public hearings: Assessing context, process and outcomes. *Canadian Journal of Urban Research, 21*(1), 106–131.

Bell, M., Fick, D., Ament, R., & Lister, N.-M. E. (2020). The use of fiber-reinforced polymers in wildlife crossing infrastructure. *Sustainability, 12*(4), 1557. https://doi.org/10.3390/su12041557.

Beudin, R. (2015). *Charrette au cul les nouvôs! Le parler des architectes.* Pierre Horay.

Bherer, L. (2006). La démocratie participative et la qualification citoyenne: À la frontière de la société civile et de l'État. *Nouvelles pratiques sociales, 18*(2), 24–38. https://doi.org/10.7202/013285ar

Bherer, L. (2010). Successful and unsuccessful participatory arrangements: Why is there a participatory movement at the local level? *Journal of Urban Affairs, 32*(3), 287–303. https://doi.org/10.1111/j.1467-9906.2010.00505.x

Bherer, L. (2011). Les relations ambiguës entre participation et politiques publiques. *Participations: Revue de sciences sociales sur la citoyenneté et la démocratie, 1*(1), 105–133. https://doi.org/10.3917/parti.001.0105

Bherer, L., & Breux, S. (2012). The diversity of public participation tools: Complementing or competing with one another? *Canadian Journal of Political Science / Revue canadienne de science politique, 45*(2), 379–403. http://doi.org/10.1017/S0008423912000376

Bherer, L., Fahmy, M., & Pinsky, M. (2015). *Professionnalisation de la participation publique: Acteurs, défis, possibilités.* Institut du Nouveau Monde.

Bird, L., & Luka, N. (2010). Arts of (dis)placement: City space and urban design in the London of *Breaking and Entering. Cinémas, 21*(1), 79–103. http://doi.org/10.7202/1005631ar

Blanchet-Cohen, N. (2014). Igniting citizen participation in creating healthy built environments: The role of community organizations. *Community Development Journal, 50*(2), 264–279. http://doi.org/10.1093/cdj/bsu031

Bornstein, L. (2010). Mega-projects, city-building and community benefits. *City, Culture and Society, 1*(4), 199–206. https://doi.org/10.1016/j.ccs.2011.01.006

Bornstein, L., & Leetmaa, K. (2015). Moving beyond indignation: Stakeholder tactics, legal tools and community benefits in large-scale redevelopment projects. *Oñati Socio-Legal Series, 5*(1), 29–50.

Bos, N., Zimmerman, A., Olson, J., Yew, J., Yerkie, J., Dahl, E., & Olson, G. (2007). From shared databases to communities of practice: A taxonomy of collaboratories. *Journal of Computer-Mediated Communication, 12*(2), 652–672. http://doi.org/10.1111/j.1083-6101.2007.00343.x

Bourdieu, P. (1977). *Outline of a theory of practice.* Cambridge University Press.

Brown, V. A. (2010). Conducting an imaginative transdisciplinary inquiry. In V. A. Brown, J. A. Harris, & J. Y. Russell (Eds.), *Tackling wicked problems through transdisciplinary imagination* (pp. 103–114). Earthscan.

Buchanan, R. (1992). Wicked problems in design thinking. *Design Issues, 8*(2), 5–21. http://www.jstor.org/stable/1511637

Burch, S., Shaw, A., Dale, A., & Robinson, J. (2014). Triggering transformative change: A development path approach to climate change response in communities. *Climate Policy, 14*(4), 467–487. http://doi.org/10.1080/14693062.2014.876342

Chupin, J.-P., Cucuzzella, C., & Helal, B. (Eds.). (2015). *Architecture competitions and the production of culture, quality and knowledge: An international inquiry.* Potential Architecture Books.

Condon, P. M. (Ed.). (1996). *Sustainable urban landscapes: The Surrey design charrette.* University of British Columbia Press.

Condon, P. M. (2008). *Design charrettes for sustainable communities.* Island Press.

Condon, P. M., & Proft, J. (Eds.). (1999). *Sustainable urban landscapes: The Brentwood design charrette.* University of British Columbia Press.

Corner, J. (1998). Operational eidetics: Forging new landscapes. *Harvard Design Magazine, 6,* 22–26. http://www.harvarddesignmagazine.org/issues/6

Corner, J. (2006). Terra fluxus. In C. Waldheim (Ed.), *The landscape urbanism reader* (pp. 21–33). Princeton Architectural Press.

Cross, N. (2007). *Designerly ways of knowing.* Birkhäuser.

Cuff, D. (1989). The social art of design at the office and the academy. *Journal of Architectural and Planning Research, 6*(3), 186–203.

Dale, A. (2001). *At the edge: Sustainable development in the 21st century.* UBC Press.

Dale, A., Robinson, J., Herbert, Y., & Shaw, A. (2013). *Climate change adaptation and mitigation: An action agenda for BC decision-makers.* Community Research Connections Report. Royal Roads University. https://www.crcresearch.org/solutions-agenda/climate-action-agenda-bc-decision-makers

Davidoff, P. (1965). Advocacy and pluralism in planning. *Journal of the American Institute of Planners, 31*(4), 331–338. https://doi.org/10.1080/01944366508978187

Daviter, F. (2019). Policy analysis in the face of complexity: What kind of knowledge to tackle wicked problems? *Public Policy and Administration, 34*(1), 62–83. http://doi.org/10.1177/0952076717733325

De Block, G. (2016). Ecological infrastructure in a critical-historical perspective: From engineering "social" territory to encoding "natural" topography. *Environment and Planning A: Economy and Space, 48*(2), 367–390. http://doi.org/10.1177/0308518X15600719

Deleuze, G., & Guattari, F. (1980). *Mille plateaux.* Editions de Minuit.

Easterling, K. (2010). Disposition and active form. In K. Stoll, S. Lloyd, & S. Allen (Eds.), *Infrastructure as architecture: Designing composite networks* (pp. 96–99). Jovis.

Easterling, K. (2014). *Extrastatecraft: The power of infrastructure space.* Verso.

Evans-Cowley, J. S., & Gough, M. Z. (2009). Evaluating new urbanist plans in post-Katrina Mississippi. *Journal of Urban Design, 14*(4), 439–461. http://doi.org/10.1080/13574800903265496

Fainstein, S. S. (1999). Can we make the cities we want? In R. A. Beauregard & S. Body-Gendrot (Eds.), *The urban moment: Cosmopolitan essays on the late-20th-century city* (pp. 249–272). Sage.

Farina, C., Kong, H., Blake, C., Newhart, M., & Luka, N. (2014). Democratic deliberation in the wild: The McGill Online Design Studio and the RegulationRoom Project. *Fordham Urban Law Journal, 41*(5), 1527–1580. https://ir.lawnet.fordham.edu/ulj/vol41/iss5/3

Fischer, F. (1993). Citizen participation and the democratization of policy expertise: From theoretical inquiry to practical cases. *Policy Sciences, 26*(3), 165–187. https://doi.org/10.1007/BF00999715

Fischer, F. (2006). Participatory governance as deliberative empowerment: The cultural politics of discursive space. *American Review of Public Administration, 36*(1), 19–4. https://doi.org/10.1177/0275074005282582

Forester, J. (1998). Creating public value in planning and urban design: The three abiding problems of negotiation, participation, and deliberation. *Urban Design International, 3*(1), 5–12. https://doi.org/10.1080/135753198350479

Forester, J. (2006). Making participation work when interests conflict: Moving from facilitating dialogue and moderating debate to mediating negotiations. *Journal of the American Planning Association, 72*(4), 447–456. http://doi.org/10.1080/01944360608976765

Forester, J. (2009). *Dealing with differences: Dramas of mediating public disputes.* Oxford University Press.

Forester, J. (2018). Deliberative planning practices without smothering intervention: A practical aesthetic view. In A. Bächtiger, J. S. Dryzek, J. J. Mansbridge, & M. Warren (Eds.), *The Oxford handbook of deliberative democracy* (pp. 595–611). Oxford University Press.

Forsyth, A. (2006). Urban centres in universities: Institutional alternatives for urban design. *Journal of Urban Design, 11*(1), 97–103. https://doi.org/10.1080/13574800500490315

Frederick, M. (2007). *101 things I learned in architecture school.* MIT Press.

Fry, G. L. A. (2001). Multifunctional landscapes: Towards transdisciplinary research. *Landscape and Urban Planning, 57,* 159–168. http://doi.org/10.1016/S0169-2046(01)00201-8

Fung, A. (2003). Recipes for public spheres: Eight institutional design choices and their consequences. *Journal of Political Philosophy, 11*(3), 338–367. https://doi.org/10.1111/1467-9760.00181

Gamon, J. A. (1991). The charrette: A technique for large groups. *Journal of Extension, 29*(3), 39. https://archives.joe.org/joe/1991fall/tt1.php

Girling, C., Kellett, R., & Johnstone, S. (2006). Informing design charrettes: Tools for participation in neighbourhood-scale planning. *Integrated Assessment Journal, 6*(4), 109–130. https://journals.lib.sfu.ca/index.php/iaj/article/view/2723

Gordon, E., & Baldwin-Philippi, J. (2014). Playful civic learning: Enabling reflection and lateral trust in game-based public participation. *International Journal of Communication, 8,* 759–786. https://ijoc.org/index.php/ijoc/article/view/2195

Grant, J. (2006). *Planning the good community: New urbanism in theory and practice.* Routledge.

Gustafson, P., & Hertting, N. (2017). Understanding participatory governance: An analysis of participants' motives for participation. *American Review of Public Administration, 47*(5), 538–549. https://doi.org/10.1177%2F0275074015626298

Hilbrandt, H. (2017). Insurgent participation: Consensus and contestation in planning the redevelopment of Berlin-Tempelhof airport. *Urban Geography, 38*(4), 537–556. https://doi.org/10.1080/02723638.2016.1168569

Horrigan, P. (2014). Rust to green: Cultivating resilience in the Rust Belt. In M. Bose, P. Horrigan, C. Doble, & S. C. Shipp (Eds.), *Community matters: Service-learning in engaged design and planning* (pp. 167–185). Routledge.

Hou, J., & Kinoshita, I. (2007). Bridging community differences through informal processes: Re-examining participatory planning in Seattle and Matsudo. *Journal of Planning Education and Research, 26,* 301–314. http://doi.org/10.1177/0739456X06297858

Hughes, H. (2017). Charrette as context and process for academic discourse in contemporary higher education: A case study. In T. Miranda & J. Herr (Eds.), *The value of academic discourse: Conversations that matter* (pp. 79–102). Rowman & Littlefield.

Irwin, T. (2015). Transition design: A proposal for a new area of design practice, study, and research. *Design and Culture, 7*(2), 229–246. http://doi.org/10.1080/17547075.2015.1051829

Jacobs, J. (1961). *The death and life of great American cities.* Vintage Books.

Katz, C. (1996). Towards minor theory. *Environment and Planning D: Society and Space, 14*(4), 487–499. http://doi.org/10.1068/d140487

Kelbaugh, D. (2011). The design charrette. In T. Banerjee & A. Loukaitou-Sideris (Eds.), *Companion to urban design* (pp. 317–328). Routledge.

Klein, J. T. (2008). Evaluation of interdisciplinary and transdisciplinary research: A literature review. *American Journal of Preventive Medicine, 35*(2S), S116–S123. https://doi.org/10.1016/j.amepre.2008.05.010

Kong, H. L., Luka, N., Cudmore, J., & Dumas, A. (2017). Deliberative democracy and digital urban design in a Canadian city: The case of the McGill Online Design Studio. In C. Prins, C. Cuijpers, P. L. Lindseth, & M. Rosina (Eds.), *Digital democracy in a globalized world* (pp. 180–200). Edward Elgar.

Krishna, A. (2001). Moving from the stock of social capital to the flow of benefits: The role of agency. *World Development, 29*(6), 925–943. http://doi.org/10.1016/S0305-750X(01)00020-1

Larsen, L. (2015). Urban climate and adaptation strategies. *Frontiers in Ecology and the Environment, 13*(9), 486–492. https://doi.org/10.1890/150103

Laurian, L. (2009). Trust in planning: Theoretical and practical considerations for participatory and deliberative planning. *Planning Theory and Practice, 10*(3), 369–391. https://doi.org/10.1080/14649350903229810

Laurian, L., & Shaw, M. M. (2009). Evaluation of public participation: The practice of certified planners. *Journal of Planning Education and Research, 28*(3), 293–309. http://doi.org/10.1177/0739456X08326532

Lehrer, U. (2011). Urban design competitions. In T. Banerjee & A. Loukaitou-Sideris (Eds.), *Companion to urban design* (pp. 304–316). Routledge.

Lennertz, W. R. (2003). The charrette as an agent for change. In R. Steuteville & P. Langdon (Eds.), *New urbanism: Comprehensive report and best practice guide* (pp. 12–18). New Urban Publications.

Lennertz, W. R., & Lutzenhiser, A. (2006). *The charrette handbook: The essential guide for accelerated, collaborative community planning*. National Charrette Institute/American Planning Association.

Lister, N.-M. (2012). Crossing the road, raising the bar: The ARC International Design Competition. *Ecological Restoration, 30*(4), 335–340. http://doi.org/10.3368/er.30.4.335

Lister, N.-M., Brocki, M., & Ament, R. (2015). Integrated adaptive design for wildlife movement under climate change. *Frontiers in Ecology and the Environment, 13*(9), 493–502. https://doi.org/10.1890/150080

Luka, M. E., & Lilley, B. (2018). The NiS+TS Psychogeographer's Table: Countering the official Halifax Explosion Archive. *Public, 29*(57), 236–249. https://doi.org/10.1386/public.29.57.236_1

Luka, N. (2018). *Civic coproduction = counterinstitutions + people: Make participation work by focusing on the possible*. The Nature of Cities. https://www.thenatureofcities.com/2018/07/09/civic-coproduction-counterinstitutions-people-make-participation-work-focusing-possible/

Luka, N., & Engle, J. (2015). Neighborhood planning for resilient and livable cities, Part 3 of 3: Montréal's Green, Active, and Healthy Neighborhoods Project. The Nature of Cities. https://www.thenatureofcities.com/2015/11/18/neighborhood-planning-for-resilient-and-livable-cities-part-3-of-3-montreals-green-active-and-healthy-neighborhoods-project/

Luka, N., & Lister, N.-M. (2000). Our place: Community ecodesign for the Great White North means re-integrating local culture and nature. *Alternatives, 26*(3), 25–30.

Macpherson, H. (2016). Walking methods in landscape research: Moving bodies, spaces of disclosure and rapport. *Landscape Research, 41*(4), 425–432. http://doi.org/10.1080/01426397.2016.1156065

Margerum, R. D. (2002). Collaborative planning: Building consensus and building a distinct model for practice. *Journal of Planning*

Education and Research, 21(3), 237–253. http://doi.org/10.1177/07394 56X0202100302

McAvoy, P. V., Driscoll, M. B., & Gramling, B. J. (2004). Integrating the environment, the economy, and community health: A community health center's initiative to link health benefits to smart growth. *American Journal of Public Health, 94*(4), 525–527. https://doi.org/10.2105/ajph.94.4.525

Miller, I. (1982). Design seminar: An urban site. *Journal of Architectural Education, 35*(4), 27–31. https://doi.org/10.1080/10464883.1982.10758303

Milroy, B. M. (2009). *Thinking planning and urbanism.* UBC Press.

Mitchell, T. D. (2008). Traditional vs. critical service-learning: Engaging the literature to differentiate two models. *Michigan Journal of Community Service Learning, 14*(2), 50–65. http://hdl.handle.net/2027/spo.3239521.0014.205

Mobjörk, M. (2010). Consulting versus participatory transdisciplinarity: A refined classification of transdisciplinary research. *Futures, 42*(8), 866–873. https://doi.org/10.1016/j.futures.2010.03.003

Moe, K. (2019). The architecture of work and the work of architecture today. *Journal of Architectural Education, 73*(2), 251–253. https://doi.org/10.1080 /10464883.2019.1633205

Moore, S. (2013). What's wrong with best practice? Questioning the typification of New Urbanism. *Urban Studies, 50*(11), 2371–2387. http://doi .org/10.1177/0042098013478231

Nassauer, J. I., & Opdam, P. (2008). Design in science: Extending the landscape ecology paradigm. *Landscape Ecology, 23*(6), 633–644. http://doi .org/10.1007/s10980-008-9226-7

Neuman, M. (2016). Teaching collaborative and interdisciplinary service-based urban design and planning studios. *Journal of Urban Design, 21*(5), 596–615. https://doi.org/10.1080/13574809.2015.1100962

Newell, R., Dale, A., Herbert, Y., Duguid, F., Foon, R., & Hough, P. (2015). Trans-disciplinary research: An academic-practitioner partnership effort on investigating the relationship between the cooperative model and sustainability. *Interdisciplinary and Multidisciplinary Journal of Social Sciences, 4*(1), 23–53. https://doi.org/10.17583/rimcis.2015.1384

Opdam, P., Nassauer, J. I., Wang, Z., Albert, C., Bentrup, G., Castella, J.-C., McAlpine, C., Liu, J., Sheppard, S., & Swaffield, S. (2013). Science for action at the local landscape scale. *Landscape Ecology, 28*(8), 1439–1445. http://doi .org/10.1007/s10980-013-9925-6

Osterwalder, A., & Pigneur, Y. (2010). *Business model generation: A handbook for visionaries, game changers, and challengers.* Wiley.

Oxford English Dictionary. (2018). (2nd ed.). Oxford University Press. www.oed.com

Polk, M., & Knutsson, P. (2008). Participation, value rationality and mutual learning in transdisciplinary knowledge production for sustainable development. *Environmental Education Research, 14*(6), 643–653. https://doi.org/10.1080/13504620802464841

Ramadier, T. (2004). Transdisciplinarity and its challenges: The case of urban studies. *Futures, 36*(4), 423–439. https://doi.org/10.1016/j.futures.2003.10.009

Rapoport, A. (1995). On the nature of design. *Practices* (3/4), 32–43.

Razzouk, R., & Shute, V. (2012). What is design thinking and why is it important? *Review of Educational Research, 82*(3), 330–348. https://doi.org/10.3102%2F0034654312457429

Rittel, H. W. J., & Webber, M. M. (1973). Dilemmas in a general theory of planning. *Policy Sciences, 4*(2), 155–169. https://doi.org/10.1007/BF01405730

Roggema, R. (2014). The design charrette. In R. Roggema (Ed.), *The design charrette: Ways to envision sustainable futures* (pp. 15–34). Springer.

Rosol, M. (2010). Public participation in post-fordist urban green space governance: The case of community gardens in Berlin. *International Journal of Urban and Regional Research, 34*(3), 548–563. https://doi.org/10.1111/j.1468-2427.2010.00968.x

Russell, A. W., Wickson, F., & Carew, A. L. (2008). Transdisciplinarity: Context, contradictions and capacity. *Futures, 40*(5), 460–472. https://doi.org/10.1016/j.futures.2007.10.005

Sanoff, H. (2000). *Community participation methods in design and planning*. Wiley.

Schluntz, R. L. (1982). Design competitions: For whose benefit now? *Journal of Architectural Education, 35*(4), 2–9. https://doi.org/10.1080/10464883.1982.10758299

Schön, D. A. (1993). Generative metaphor: A perspective on problem-setting in social policy. In A. Ortony (Ed.), *Metaphor and thought* (2nd ed., pp. 137–163). Cambridge University Press.

Schön, D. A., & Rein, M. (1994). *Frame reflection: Toward the resolution of intractable policy controversies*. Basic Books.

Schröder, T. (1993). Die Chance des Potsdamer Platzes? ein Platz im Spiegel der Wettbewerbe. In H. Mauter (Ed.), *Der Potsdamer Platz: Eine Geschichte in Wort und Bild* (pp. 151–163). Verlag Dirk Nishen.

Shaw, R. (2002). The International Building Exhibition (IBA) Emscher Park, Germany: A model for sustainable restructuring? *European Planning Studies, 10*(1), 77–97. https://doi.org/10.1080/09654310120099272

Shotwell, A. (2016). *Against purity: Living ethically in compromised times*. University of Minnesota Press.

Sorensen, A. (2011). Uneven processes of institutional change: Path dependence, scale and the contested regulation of urban development in Japan. *International Journal of Urban and Regional Research, 35*(4), 712–734. https://doi.org/10.1111/j.1468-2427.2010.00975.x

Stevens, G. (1998). *The favored circle: The social foundations of architectural distinction*. MIT Press.
Sutton, S. E., & Kemp, S. P. (2002). Children as partners in neighborhood placemaking: Lessons from intergenerational design charrettes. *Journal of Environmental Psychology*, 22(1–2), 171–189. http://doi.org/10.1006/jevp.2001.0251
Sutton, S. E., & Kemp, S. P. (2006). Integrating social science and design inquiry through interdisciplinary design charrettes: An approach to participatory community problem-solving. *American Journal of Community Psychology*, 38(1–2), 51–62. http://doi.org/10.1007/s10464-006-9065-0
Swain, C., & Tait, M. (2007). The crisis of trust and planning. *Planning Theory and Practice*, 8(2), 229–247. https://doi.org/10.1080/14649350701324458
Swyngedouw, E., & Moulaert, F. (2010). Socially innovative projects, governance dynamics and urban change: Between state and self-organisation. In F. Moulaert, F. Martinelli, E. Swyngedouw, & S. González (Eds.), *Can neighbourhoods save the city? Community development and social innovation* (pp. 219–234). Routledge.
Termeer, C. J. A. M., Dewulf, A., Breeman, G., & Stiller, S. J. (2015). Governance capabilities for dealing wisely with wicked problems. *Administration and Society*, 47(6), 680–710. https://doi.org/10.1177%2F0095399712469195
Thompson, M. A., Owen, S., Lindsay, J. M., Leonard, G. S., & Cronin, S. J. (2017). Scientist and stakeholder perspectives of transdisciplinary research: Early attitudes, expectations, and tensions. *Environmental Science and Policy*, 74, 30–39. https://doi.org/10.1016/j.envsci.2017.04.006
Toker, U. (2012). *Making community design work: A guide for planners*. Taylor & Francis.
Tufte, E. R. (2006). *Beautiful evidence*. Graphics Press.
Van Herzele, A. (2004). Local knowledge in action: Valuing nonprofessional reasoning in the planning process. *Journal of Planning Education and Research*, 24(2), 197–212. http://doi.org/10.1177/0739456X04267723
Van Herzele, A., & van Woerkum, C. M. J. (2008). Local knowledge in visually mediated practice. *Journal of Planning Education and Research*, 27(4), 444–455. http://doi.org/10.1177/0739456X08315890
Van Neste, S. L., & Martin, D. G. (2018). Place-framing against automobility in Montreal. *Transactions of the Institute of British Geographers*, 43(1), 47–60. https://doi.org/10.1111/tran.12198
Vardouli, T. (2016). User design: Constructions of the "user" in the history of design research. Proceedings of the Design Research Society 50th Anniversary Conference. Design Research Society. https://www.drs2016.org/262
Waldheim, C. (2016). *Landscape as urbanism: A general theory*. Princeton University Press.

Walters, D. (2007). *Designing community: Charrettes, master plans and form-based codes*. Elsevier/Architectural Press.

Wates, N. (2008). *The community planning event manual: How to use collaborative planning and urban design events to improve your environment*. Earthscan.

Wates, N. (2014). *The community planning handbook: How people can shape their cities, towns, and villages in any part of the world* (2nd ed.). Earthscan/Urban Design Group/The Prince's Foundation.

Weinstock, D. (2013). On the possibility of principled moral compromise. *Critical Review of International Social and Political Philosophy, 16*(4), 537–556. https://doi.org/10.1080/13698230.2013.810392

Westley, F., & McGowan, K. (2020). Design thinking, wicked problems, messy plans. In C. Reed & N.-M. Lister (Eds.), *Projective ecologies* (2nd ed., pp. 294–315). ACTAR.

Wickson, F., Carew, A. L., & Russell, A. W. (2006). Transdisciplinary research: Characteristics, quandaries and quality. *Futures, 38*(9), 1046–1059. https://doi.org/10.1016/j.futures.2006.02.011

Willis, D. (2010). Are charrettes old school? *Harvard Design Magazine, 33*(2). http://www.harvarddesignmagazine.org/issues/33/are-charrettes-old-school

Zeisel, J. (2006). *Inquiry by design: Environment/behavior/neuroscience in architecture, interiors, landscape, and planning* (Rev. ed.). Norton.

Zukin, C., Keeter, S., Andolina, M., Jenkins, K., & Delli Carpini, M. X. (2006). *A new engagement? Political participation, civic life, and the changing American citizen*. Oxford University Press.

Zukin, S. (1988). The postmodern debate over urban form. *Theory, Culture and Society, 5*(2–3), 431–446. https://doi.org/10.1177%2F0263276488005002013

Zukin, S. (2010). *Naked city: The death and life of authentic urban places*. Oxford University Press.

Afterword: Thinking Through Transdisciplinarity in Urban Sustainability

TANYA MONFORTE

The most vulnerable persons in society are already bearing the greatest burden of climate change. According to the special report on global warming of 1.5°C (IPCC, 2018), if there is not dramatic change, warming of the planet will create more poverty and social disadvantage, and urban settings will be hit the hardest. A small window is still open for the societal shift needed to halt climate change. "Without societal transformation and rapid implementation of ambitious greenhouse gas reduction measures, pathways to limiting warming to 1.5°C and achieving sustainable development will be exceedingly difficult, if not impossible, to achieve" (p. 448). The urgency is especially evident in Canada, where warming is occurring at double the rate of the rest of the world, already having surpassed the 1.5°C mark, hitting 1.7°C (Canada, 2019). Warming, habitat loss, reductions in biodiversity, pollution, and resource depletion are crises we appear to be ill equipped to correct with current tools and economic frameworks. Although in the past Canada has been considered a global leader in the fight against climate change (Buxton, 1992), that merely meant that Canada had done marginally better than other countries as we collectively march towards the tipping point. In recent years Canada has lost its role as a global leader in sustainability. That being said, the Montreal climate strike of September 27, 2019, was one of the largest in the world and reportedly the largest protest in Quebec history. With popular support for climate action, Montreal's Mayor has announced an ambitious plan to reduce the city's emissions by 55 per cent within ten years (CBC, 2019). While Quebec is known for radical, participatory movements, Canada and Quebec together, despite deep differences, have managed to peacefully find political common ground, including in contentious areas such as climate change. Now more than ever we need a radical political shift mediated through democratic processes. This collection brings law, governance, and social movements together in novel ways for urban studies that incorporate radical movements grounded in a democratic framework.

Rethinking how to make the environment liveable for present and future generations and finding new approaches to complex problems in a manner that is economically and socially inclusive is nothing less than a societal imperative. Social mobilization and responsive state action at different levels of government are part of the social imperative.

This collection deals with hard problems facing societies seeking sustainability. The chapters in this volume touch on problems of the political and legal order while working with the challenge of managing competing interests and conflicting political positions. This collection attempts to bridge a gap in the academic literature by gathering together authors from across Canada who work in diverse disciplines, including law, and by addressing urban sustainability issues at the granular level. Some of the chapters offer practical suggestions, including how to garner institutional support for social movements. Others highlight collective action problems for which we do not yet have good solutions.

This is often the place in an edited collection where the tensions and conflicts between the chapters are resolved and differing uses of terminology get settled. This settling of language would be a mistake, however, since we should seriously consider the role of the reader as a participant in the production of knowledge. This collection should be read not only for the individual chapters in isolation, nor as a single coherent product for consumption, but rather as a collection that is in conversation with the reader. This collection brings together different disciplinary and political perspectives and allows productive tensions to remain, for the benefit of both the authors and the readers. For this reason, I don't want to force an artificial resolution of every point of contention about, for example, what "sustainability" entails. Authors in this collection and readers almost certainly disagree, for example, whether or not sustainability implies a model of sustainable growth or if economic degrowth is in fact a requirement at this moment. Thus, the very concept of urban sustainability is left unsettled. The authors approach the topic of urban sustainability such that in one chapter, municipalities are read as underpinning the "urban," while others approach the subject geographically or through its legal structure. Different disciplines have distinctive ways of speaking about the methods associated with urban sustainability. Should we discuss "consultation" or "citizen engagement" or simply "participation"? And do we use the terms "citizens" or "residents" or "stakeholders," as each of these also has different built-in political assumptions? Similarly, while innovation has become an important concept in discussions about sustainability (Newig et al., 2019), its meanings and political implications differ, depending on whether authors are speaking about social or technological or political innovation. Many of the conflicting uses of terms and vocabularies reflect disciplinary distinctions, and, importantly, some political divisions that often are embedded if not masked by disciplinary conventions.

In my own field of law, interdisciplinary work is increasingly the norm and I am myself situated in the interdisciplinary cross-section of law and society. Law has become a "law and ———" kind of discipline, that is to say it is a discipline in which most legal scholars already approach law interdisciplinarily as "law and economics" or "law and society" (Schleicher, 2016). This interdisciplinarity is often based in two fields of specialization and accreditation. As a professional degree, the J.D. is a terminal degree, and it is often complimented by a PhD or an M.A. in another field such as sociology, economics, or political science. The legal scholar thus is often an "expert" in more than one field.

This way of doing interdisciplinary work has been a question of debate in the field of law for urban law scholars. David Schleicher, who writes on local government law out of Yale Law School, has claimed that the field of local government law has failed to keep up with contemporary social sciences and is in his words "outdated," as there is little to no engagement with agglomeration economics and positive political science (Schleicher, 2016). In this argument, being current and relevant assumes a particular bibliography, a kind of hyper-specialization with epistemic assumptions built in, but it is presented as disciplinary necessity. My intention is to explain why this position channels vocabularies and thinking in very predictable ways and can create a roadblock for us as we face real crisis. It is for this reason that I want to end on a methodological reflection.

From my point of view as a legal scholar, the way we combine disciplines when we are "thinking of ourselves as urban law scholars," as David Schleicher writes, is itself a political choice. The kind of approach to the subject we take – be it a quantitative approach to law and the city as Schleicher endorses, or a critical approach, or both – can change the kind of knowledge produced. This collection breaches disciplinary boundaries by including diverse political/disciplinary assumptions. Breaching disciplinary boundaries is one manner of working through seemingly intractable social problems, seeing them in new ways, and even arriving at novel approaches to concrete problems we face in urban settings. We are, in other words, breaching politically determined boundaries – sometimes masked as disciplinary boundaries – in order to draw out points of commonality, contention, and conversation, and hopefully to incite collective action. Facing environmental crises does not require giving up our convictions, but it does necessitate expanding conversations beyond the limited groupings of persons with whom we already agree.

This volume does not aim to speak to or through any particular discipline or vocabulary but approaches urban sustainability through several different lenses and vocabularies and with different bibliographies behind them. This methodological, political, and disciplinary eclecticism is a distinct strength of the volume and reflects the variety of approaches and perspectives needed to address the complexity of urban issues. This volume does not aim to provide a technocratic list of "best practices," but instead attempts to think through the relevant social problems in different ways and to allow tensions to contribute to the production of knowledge and politics. Importantly the authors come from theoretically and politically different positions. This is a volume that includes literature on *both* social movements and governance; these are topics that tend to become siloed for political reasons as much as disciplinary ones, and this collection aims to break down these siloes.

As a whole, this collection could be read as employing many of the conventions of transdisciplinarity. Widespread in environmental studies, transdisciplinarity (or TD) is an approach that naturally lends itself to a volume on urban sustainability and to the kinds of questions that have been addressed throughout this book. Transdisciplinary approaches are understood as "a form of research that is driven by the need to solve problems of the life-world" (Hirsch Hadorn et al., 2008, p. 20). It is particularly useful for addressing complex social problems that defy specialization and cannot be characterized simply as, for example, an environmental, economic, or legal problem because they necessarily implicate many disciplines. It is sometimes described more as an approach than a method and is often characterized as delineating the disposition of the researcher towards certain principles (Bunders et al., 2015, p. 27). In this afterword, I treat it as an approach that has certain methods attached to it (such as participatory methods), but that also engages a researcher's sensibilities. Despite variations in the transdisciplinary approach over time and across fields, four overlapping features constitute the core characteristics of a TD approach: (1) it is often participatory; (2) it transcends and integrates disciplinary paradigms; (3) it "focuses on real-world problems"; and (4) ultimately it looks to produce knowledge beyond disciplinary boundaries (p. 27; Pohls & Hirsch Hadorn, 2007).

There is no single formulaic way to do TD research. Many writers who specialize in transdisciplinarity agree that flexibility in the approach is an additional key component (Mobjörk, 2010; Osborne, 2015). Depending on the types of issues addressed, there will likely be different styles. It is a bit paradoxical to say there are different disciplinary conventions of transdisciplinarity, but it does appear that as an approach, it is in fact developing slightly differently in separate fields or disciplines, perhaps in part because classes of problems are often field specific.[1] Transdisciplinarity has gained a strong foothold in environmental studies as well as in public health – fields in which the complexity of concrete problems often requires a variety of specializations (Pennington et al., 2013; Polk & Knutsson 2008). The ways of integrating knowledge from different disciplines or fields varies, depending on the kind of problem that is being addressed.

However, I would like to depart from the social science variation of the approach that, as we have seen above, veers towards methodological rigidity in the urban law context. Instead, I emphasize the *trans* element of transdisciplinarity in order to explore how the concept can be a source of innovation. In this view TD can involve speaking from one's expert position in a discipline to other experts, but it can also mean talking through disciplines or against one's own discipline in the sense of working against disciplinary dogmas that create seemingly necessary methodological and political choices. There is a kind of fluidity in this approach to TD that allows for discipline-hopping or resistance to strict disciplinary conventions. This approach bends disciplines, in the sense that it allows for breaking boundaries and not only crossing boundaries. It is flexible and reflexive. It challenges the naturalness of the contours and boundaries of disciplines and the givenness of their political commitments. There is something deeply innovative about transdisciplinarity when read this way, as it "is not the bridging of existing disciplines; it is their transcendence by a new epistemology" (Macdonald, 2000, p. 69). That being said, the mechanics of how transdisciplinarity gives rise to innovation needs elaboration.[2]

If we think of this collection as a "trans" collection, then there are a few benefits of transdisciplinarity to be highlighted. Coming as it does at the end of a collection, this methodological reflection offers another

---

1 There are many ways to categorize the different strands in transdisciplinary research. I have chosen to address them as field- and question-driven. However, a very helpful analysis groups them into discourses that partially map onto fields. This analysis comes from Peter Osborne (2015).
2 Innovation as a new way of thinking can also mean arriving at very old ideas through a new set of knowledge practices. In their best form, sustainability practices would be, in my mind, a new way to state what was already known on this continent long before the land was colonized. See Leanne Simpson (2008, p. 37), who writes on pre-colonial treaty relationships and the early principles that encompassed what we need to know for sustainability practices:

Afterword: Thinking Through Transdisciplinarity in Urban Sustainability

way to think of how transdisciplinarity can work in collective projects. As we transcend binaries or siloes in thinking and take steps in the direction of new vocabularies, there may be some messiness and some contradictions, but this movement is its own form of innovation. Epistemic innovation can break path dependence and so may be one avenue towards addressing crises.

Hannah Arendt, Saskia Sassen, and Jane Jacobs have brought about innovative ways of thinking through social problems, changed paradigms, and broke through disciplinary conventions, in part by being unconventional. These three will serve as examples of how to "acquire genuinely new knowledge (previously unknown to anyone or anything who could otherwise transfer it to you)" (Nickles, 2003, p. 64). Examining these authors allows me to open up the epistemic innovation problem, because they affirm and challenge some of the ways transdisciplinarity is currently being conceptualized. I argue that there are many ways to garner the benefits of transdisciplinarity. The authors show us that TD may be less about a rigid method that must be followed and more like a sensibility that an author or group of authors have or can use when researching or reading. The conventions of this approach should be flexible. Rather than simply debating the proper definition of transdisciplinarity, I will focus on pragmatic qualities of transdisciplinarity that are important for the complex social, economic, and environmental problems we face, and I will address some ways in which we can approach urban sustainability for better outcomes. These three authors individually employ methods that capture some of the positive attributes of transdisciplinarity without formal affiliation with the approach. They represent a certain comfort with discomfort and an ability to change paradigms of thought.

Arendt, Sassen, and Jacobs highlight aspects of transdisciplinarity that challenge disciplinary boundaries while maintaining a focused interest in social inclusion and political transformation. Their methodological choices have affinities with TD, and lessons we can draw from them may challenge the ways we decide to approach knowledge production that

---

> Nishnaabeg environmental ethics dictated that individuals could only take as much as they needed, that they must share everything following Nishnaabeg redistribution of wealth customs, and no part of the animal could be wasted. These ethics combined with their extensive knowledge of the natural environment, including its physical features, animal behavior, animal populations, weather, and ecological interactions ensured that there would be plenty of food to sustain both parties in the future. Decisions about use of resources were made for the long term. Nishnaabeg custom required decision makers to consider the impact of their decisions on all the plant and animal nations, in addition to the next seven generations of Nishnaabeg.

crosses boundaries. These three authors have been exalted and criticized for their unconventional methodological choices and rebellious sensibilities. Each has been positioned at the centre of and as an outsider to her respective "discipline." What they have in common, apart from the spark that has been called brilliance, is that they are all creative, innovative, and engaged thinkers who dig into social problems, rebelling against disciplinary conventions. Jane Jacobs and Saskia Sassen are more obvious choices for discussing the methodology of urban sustainability, since Jacobs made her career thinking through problems as an activist and Sassen has published directly on this topic. Arendt is a less obvious choice. She, however, wrote extensively on social movements. Pieces such as *Reflections on Little Rock* remain challenging texts that focus on the intersections of law, social activism, and politics.

Five points that can be drawn out from Jacobs the urban author, Arendt the non-philosopher refugee, and Sassen the engaged migrant that are useful in thinking through transdisciplinarity; each point is characteristic both of their work and of transdisciplinarity, and each has clear benefits for social enquiry. The five characteristics are looking with an undisciplined eye, resisting the fetishization of expertise, decompartmentalizing systems of knowledge, including the excluded, and using accessible language. While the literature on cities is often eclectic, and people working in urban studies are frequently used to reading literature from specializations and disciplines different from their own, we should ask if there is something that makes a transdisciplinary, rather than multidisciplinary, approach to these activities special.

1. Looking with an Undisciplined Eye

The idea of transcending disciplinary boundaries can become almost formulaic in transdisciplinary work if it only means having a string of co-authors from different disciplines who synthesize the information examined in an article into a coherent whole. Instead, transcending one's discipline can be something like what Picasso suggested was his method of trying to paint like a child. There are benefits to be gained from looking at the world with an undisciplined eye – to look without following rigid rules that come from being "trained." Sassen, Arendt, and Jacobs have been celebrated as original thinkers, and perhaps part of that originality comes from methodological choices, either conscious or unconscious. They all observed and described social phenomena in novel ways that revealed an undisciplined eye in the best sense of the term. Ignoring disciplinary conventions is a way of *resisting* predetermined disciplinary conclusions.

Transdisciplinarity can involve resistance to disciplinary conventions (Osborne, 2015). As a form of resistance, it could be contrasted with interdisciplinarity. Interdisciplinary approaches suggest speaking from one's discipline to another discipline. It involves knowledge produced within disciplinary conventions, even if they may be the disciplinary conventions of another discipline. As an example, an interdisciplinary sociologist may look at a set of laws, through the conventions of sociology. Transdisciplinarity, by contrast, requires an openness to suspending disciplinary conventions for some particular end.

Resistance to disciplinary conventions can emerge from an absence of training. Jane Jacobs was not necessarily resisting a disciplinary background, since she came into her field without formal training. It was the power of her ideas and her observations that made her first book, *The Death and Life of Great American Cities*, such a force (Jacobs, 1961). Saskia Sassen, on the other hand, has spoken about her approach to social issues as "before method" and as intentionally unconstrained (Sassen, 2013a). She consciously rejects taking some categories as givens, and instead interrogates the very foundations of her discipline. She writes,

> Such an experimental rumination, I find, requires the freedom to suspend, even if temporarily, method and its disciplining of the what, the how, and the why of an inquiry. I need to engage in what I think of as analytic tactics – the freedom to position myself in whatever ways I want/need vis-à-vis the object of study. I think of this as the space "before method." (p. 79)

Sassen's approach, like transdisciplinarity, is reflexive and needs-based. Its form depends on the object of study or the particular question of enquiry. Additionally, Sassen's "need to destabilize all major categories" allows for innovative insights that methodological conventions would otherwise preclude (Sassen, 2013b). The destabilizing of categories that are taken as givens in any particular discipline can be challenging or even destabilizing for readers within that discipline.

Likewise, Hannah Arendt's unique approach to social enquiry often defied disciplinary categorization altogether and resisted the established categories of analysis of the time. As an example, Arendt's use of the term "totalitarianism" was controversial when *The Origins of Totalitarianism* first came out (Arendt, 1951). Some commentators puzzled over the methodological choices in Arendt's book (Voegelin, 1953). Before Arendt had done it, it was unorthodox to put Stalin and Hitler in the same frame to analyse totalitarianism, and she was harshly criticized for it. Seyla Benhabib explains,

Arendt is searching for the "elements" of totalitarianism; for those currents of thought, political events and outlooks, incidents and institutions, that once the "imagination of history" gathered them together in the present, reveal an altogether different meaning than what they stood for in the original context. (Benhabib, 2003, p. 64)

In other words, she challenged received knowledge by looking at a granular level for social explanation.

A reflexive openness to being driven by the enquiry along with a resistance to constraint by disciplinary conventions are characteristics of a transdisciplinary approach. These characteristics can be translated into a concrete advantage: seeing a social problem differently can open up new approaches and potentially new solutions to a given social problem. This advantage is an epistemic point, derived from resisting the disciplining of method. Durand Folco's championing of "citizen-centric rebel cities" exemplifies a new way to conceptualize the city by looking at how certain vocabularies fit into different socially determined power structures. For urban sustainability there are many categories that appear to be foundational, such as sustainability, innovation, and rights. Rather than taking them as givens, we can challenge the processes and the conceptual foundations of disciplines; by entertaining contradictory approaches to foundational concepts, we can in turn open the field to new possibilities.

2. Resisting the Fetishization of Expertise

It cannot be that only lawyers can talk about laws, only economists can talk about markets, and only sociologists or anthropologists can talk about society. We certainly gain something from specialization, but we lose something as well when we require that observers of the places we live in hold degrees based in part on specialized vocabularies and bibliographies. This kind of requirement necessarily involves an exclusionary practice. We also lose something if, by adopting this kind of practice, we fail to see the radical potential of discipline-bending as an analytical tool.

Jane Jacobs, it has been very frequently noted, was self-taught, publishing her books without having been formally trained as an economist, sociologist, or urban anything (Laurence, 2007). Not only did she refuse to be called an expert after having written the most influential book on city planning in the twentieth century, she claimed that expert knowledge was often "dangerous" since "traffic engineers who know off the top of their heads how many vehicles any type of road can carry in an hour are never inclined to ask whether the road ought to be there in the first place" (Goldberger, 2006, p. 123). When Jacobs insisted on being referred to as an *author*, rather than an expert, she was making a rather radical political statement about who has the authority to speak about the city. It was the quality of the observation, not the strict following of disciplinary conventions or her status as an expert, that made her work resonate with so many.

The problem of fetishizing expertise is multifaceted. Experts do not simply speak from a place of authority, but they also carve out a space for their expertise and in that sense constitute the field of their expertise (Kennedy, 2018, p. 4). To speak about a social problem and pronounce a solution from the expert position of a lawyer may convert that problem into a legal or technical problem, and nothing more. It may be a way to try to end discussion or debate on a political struggle by asserting an expert opinion. When expert knowledge is required for authority to speak intelligently about a topic, the implication is that unless you know all the proper ways to use a vocabulary you cannot know the right solution to a problem. The political process becomes a black box of insulated technical knowledge. In opposition to this tendency of allowing expertise to determine merit, Richard Shearmur's chapter on municipal innovation challenges the expert and often disciplinary-specific presumption that innovation is necessarily good. He instead proposes that innovation is itself political and therefore should be evaluated according to complex and "competing value systems."

Transdisciplinarity should not only be about speaking against one's own discipline but also about speaking on a topic that has been claimed by the disciplinary expertise of another. Sassen, Arendt, and Jacobs have also been derided for speaking on topics and themes that are not "their own." It was not until twenty years after Arendt had written *The Origins of Totalitarianism* that her historical approach was recognized as a "distinctive" "hybrid" take on history (Krieger, 1976, p. 675). In 1952 Arendt's training was seen as a stain on her ability to speak on history, as demonstrated by one reviewer who wrote,

> Disregard of the concept of historical continuity may well be accountable for a persistent tendency toward overstatement. Trained in philosophy and theology, the author is inclined to cast judgements in absolute and categorical forms. In part brilliant and suggestive, the work as a whole conveys an impression of miscarriage. (Van Duzer, 1952, p. 934)

In a similar vein, John Friedmann's review of Jacobs's *The Economy of Cities* was particularly harsh. He criticized her writing on the economy for, among other things, lacking scientific facts; he characterized her approach as more philosophical than scientific, and he castigated her failure to cite the relevant literature, which, he claimed, she did not even appear to be aware of (Friedmann, 1970). The critique of Jacobs for failing to cite a literature she entirely departs from is perhaps less an instance of myopia and more an assertion of a right to a subject area. The assertion of expertise is a way to claim turf so that, for example, only lawyers can opine on problems with legal dimensions, in accordance with legal conventions. A difficulty with this kind of assertion is that it will constrain the horizon of possible solutions. Just because someone is a legal expert does not mean that application of legal expertise to a given social problem will yield a meaningful solution.

To address a concrete problem using a new vocabulary and lens, especially when one does not hold the title of "expert," can be an act of bravery (that is, of course, when it is not merely hubris). In the area of urban sustainability, the complexity of the problems we face is evidenced by the number of intertwined areas that are touched: law, the environment, markets, social relations, and health, to name only a few. Resisting the fetishization of expertise deepens democratic processes because this act of resistance asserts that, while no single citizen is an expert in all fields, no field should be treated as a sacred cow, so that only a narrowly prescribed mode of talking is allowed for us to weigh in or make decisions.

3. Decompartmentalizing Systems of Knowledge

The transdisciplinary approach can also be seen as a reaction to increasingly specialized areas of knowledge. This observation comes close to the previous point about fetishizing expertise, but there is a difference. Expertise is about an individual's authority to speak, whereas the compartmentalization of knowledge is a larger social phenomenon. The compartmentalization of knowledge is at once an epistemic phenomenon and an ontological one. It occurs at different levels of society as social institutions and the knowledge that governs them complexifies with differentiated logics. The drive to decompartmentalize knowledge is a pragmatic one, in that it aims to address social problems that may arise in part as the result of social differentiation.

Joske Bunders and co-authors write on the challenges of transdisciplinary research in relation to sustainability and explain why TD approaches are useful in this field: "In our experience, the main obstacle to a sustainability culture is the compartmentalization of our society and of our institutions in particular, each with its own goals and incentive structures" (Bunders et al., 2015, p. 24). Research on sustainability often employs transdisciplinary approaches because even if it cannot change social structures or institutions directly, it can change the way knowledge is produced about society.

As a form of resistance, it may be useful at times to work towards de-complexifying. This is one approach Sassen takes. In *Expulsions*, she argues that complexity often leads to kinds of brutality. Sassen provides a picture of different social and economic relationships to demonstrate the way complexity produces brutality and then explains the methods that allow her to make certain connections. As she observes,

> The specialization of research, knowledge, and interpretation each with its own canons and methods for protecting boundaries and meanings, does not always help in this effort of detecting subterranean trends that cut across our familiar distinctions. But specialization does give us detailed knowledge about specifics, bringing us back to basics that can be compared with one another. Rather than giving meaning to facts by processing them upward through theorization, I do the opposite, bringing them down to their most basic elements in an effort to de-theorize them. (Sassen, 2014, p. 6)

A move against modernist epistemologies or ways of knowing, de-complexifying is a move to speak in a less differentiated conversation, with – and not simply to – others without specialist knowledge. Again, the pragmatic needs-based approach means that the kind of problem will determine the precise approach. The social costs of specialization at times may outweigh the intellectual benefits. Sassen suggests that it is possible to aim towards having conversations and learning processes that are informed by specializations but are not determined by them.

Decompartmentalizing knowledge also permits seemingly contradictory approaches to come together in a single project. Transdisciplinarity is a pragmatic approach to social problems that allows multiple theoretical, disciplinary, and methodological approaches to be employed simultaneously. In cooperative approaches and with different specializations working together, it is not just possible but also likely to have contradictory logics employed simultaneously. Similar to situations in which incompatible theoretical assumptions are used in the natural sciences, when a needs-based approach is adopted, the theory that has the best explanatory power or political possibilities should be applied to any facet of a given problem. Hannah Arendt, for example, references Edmund Burke and identifies Rosa Luxemburg as a significant influence. A lesson one could draw from her eclecticism is that following a coherent bibliography won't necessarily produce the most innovative forms of knowledge. Of course, this eclecticism can lead to misinterpretation of particular sources or disciplines. For a problem-oriented approach, however, there is no requirement that there be a single theoretical orientation or even theoretical coherence for an overall approach to be useful. It is not an interesting or even meaningful question to ask, for example, if someone is an Einsteinian or a Newtonian. Despite having contradictory assumptions, a single scientist can use both Einsteinian and Newtonian physics in order to do different tasks while solving a single problem. Totalizing identification with a theoretical approach is a way to signal purity, but also to compartmentalize knowledge.

The advantages of decompartmentalizing knowledge may be in revealing some social relations or in making what was previously unseen apparent, as Sassen writes. Decompartmentalizing is an ideology that can be applied to governance and make institutions more accessible to everyone.

4. Including the Excluded

Inclusivity is a core principle within transdisciplinary approaches, and participatory methods have become the gold standard of inclusion in TD research, according to some theories (Bunders et al., 2015). Inclusion is one of the most important themes discussed in this collection, as every single chapter develops some line about what inclusion means. However, this collection highlights challenges and dangers of certain participatory methods. Participation and inclusion should not be substituted for one another, as they are not coterminous. Participation may be a possible method to get to inclusion, but inclusion is ultimately the goal. In each of their respective chapters, Kong, Curran, and Flynn point out forms of participatory politics that do not promote social inclusion. Some participatory methods, especially those that pay no attention to differences in power structures, may not be inclusive, and it is possible to do inclusive research that does not necessarily use participatory methods.

Participation is considered important for research in sustainability studies and in transdisciplinary research. Participation has become a dominant feature, if not the dominant one, of transdisciplinary approaches in the social sciences.[3] What the participation of relevant subjects entails is not, however, a straightforward matter for any research methodology, and the complexity of the topic has created a cottage industry of writing within sustainability studies. Even the kinds of actors to

---

3 As an example of this point, see the introductory statement on the foundation of the journal *Conjunctions: Transdisciplinary Journal of Cultural Participation* (Reestorff et al., 2021).

be included varies in transdisciplinary approaches; some include "stakeholders" while others include "civil society actors" (Fritz, 2016). Transdisciplinary research needs to elaborate on the "who," "when," and "how many" of participatory research.

Collaboration, co-production, and participation are often used interchangeably in relation to TD approaches. Nonetheless, within the short history of transdisciplinary research, researchers have begun to identify divergent tendencies and lines of thought about these terms (Osborne, 2015). There seems to be a divergence in fields based on the kinds of problems addressed. In the sciences, the emphasis is often placed on the benefits of collaboration for research among experts. In the social sciences, there is greater emphasis on participatory research with stakeholders. Participation differs from collaboration in that the former highlights and attempts to correct for the exclusionary dimensions of knowledge production, while the latter simply addresses knowledge specialization. The issue of stakeholder participation as a form of social inclusion is emerging as a particularly tricky issue within TD approaches in the social sciences (Fritz, 2016).

Participatory methods can fundamentally change how knowledge is produced and often what knowledge is produced. However, there are many different ways of constructing participatory work, ranging from co-production to consultative participation (Mobjörk, 2010). Participatory approaches are often proposed when researchers using traditional methods do not have the necessary experience or knowledge, and a concrete or measurable outcome is expected (Lilja & Bellon, 2008). It has been noted that a transdisciplinary approach that includes stakeholders as subjects or co-producers of research can inform the process of answering a research question, and even change the questions researchers ask (Simon et al., 2018, p. 589). Co-production of knowledge is useful when researchers are unfamiliar with a particular context, such as when they focus on rural development without having lived in a rural community or possessing knowledge of the context. But co-production can also be useful where researchers are embedded in communities, as is often the case in urban studies. Here co-producers benefit from the range of perspectives included in collaboration (Simon et al., 2018). In short, making those with the most at stake subjects and not merely objects of research has many advantages.

In urban planning, community participation has long been valued, and Jane Jacobs herself championed it (Ellis et al., 2015). She famously wrote, "Cities have the capability of providing something for everybody, only because, and only when, they are created by everybody" (Jacobs, 1961, p. 238). Oftentimes transdisciplinary research is considered research that is conducted outside the university as a distinctive mode of knowledge production (Polk & Knutsson, 2008, p. 644). Livia Fritz has argued that in the field of urban planning, participatory approaches transfer power from the haves to the have-nots and are not therefore designed only for the benefit of "knowledge production or different epistemes in research processes" (Fritz, 2016, p. 116). She suggests that the theories of power that researchers use for thinking about participatory methods do not lead them to ask why a specific participatory method is used in a given context.

Research on urban sustainability potentially touches on as many different stakeholder groups as there are groups in a city; hence research that involves participatory inclusion often produces exclusions (Agger, 2012, p. 30; see also Kong, Curran, and Flynn in this volume). There are methodological questions about what collaboration means in the context of research involving particularly vulnerable and marginalized groups and individuals (Agarwal, 2001; Agger, 2012). To avoid tokenism, the research team must include a comprehensive group of stakeholders that avoids the traps of choosing the most privileged of any group. It may therefore be prohibitively costly to conduct meaningful participatory research. The chapter in this collection on transdisciplinary charrettes superbly illustrates how the choice of research problem, method, and process will influence the way inclusion should be done in a TD approach (see Luka et al. in this volume).

However, it is too quick to say simply that participatory methods are a necessary characteristic of transdisciplinary research. Making stakeholder involvement a methodological requirement of TD approaches (Bunders et al., 2015) can constitute a possible methodological blind spot and limitation. The limitation is especially evident as transdisciplinarity becomes synonymous with mobilization, and the concept functions as a prerequisite for receiving research funding (Beland Lindahl & Westholm, 2014, p. 155; Mobjörk, 2010). Boniface Kiteme and Urs Wiesman argue that "stakeholder" involvement should be "requirement driven" so that who is involved in the research will depend on the needs of the research (Kiteme & Wiesmann, 2008, p. 75). This pragmatic and reflexive requirement-driven approach to participation seems to better meet the transdisciplinary sensibility than would an approach that makes participation a rigid methodological requirement. Ultimately, the transdisciplinary sensibility should substantively embrace inclusive research, and stakeholder participation should be regarded as an ideal when conditions allow for it to be effective.

Arendt and Sassen demonstrate an inclusive sensibility in their overlapping research on systemic exclusion, particularly the exclusion of migrants, refugees, or stateless persons. Both have written on the complex

Afterword: Thinking Through Transdisciplinarity in Urban Sustainability 209

ways that governance through law and the economy can make vulnerable persons extraneous to place and to communities (Arendt, 1951; Sassen, 2014). As it happens, Arendt's work on refugees was informed by her own experience as a refugee (Arendt, 1994). Sassen's work on migrants is similarly informed by her own experience as a migrant. Making the situated position of the author apparent and looking for a diversity of subjects is a relatively recent convention for ensuring epistemic inclusion. However, this is different from stakeholder participation and does not suffer from many of the pitfalls of stakeholder participation. There is no requirement that Arendt represent the collective interests of refugees. She writes from a situated position informed by her experiences; it is partial and limited, yet even a fragment of partial knowledge can ring true. This is a different way of doing inclusivity. It values substantive inclusion and recognizes that participatory practices can be socially exclusionary. In their chapter, Manaugh and Dreszer address social movements in Montreal that advocate for pedestrian space, and provide concrete lessons for popular mobilization. The chapter could be read as being addressed to the very people they are writing about. Much like Arendt and Sassen's situated approach, writing *to* people is a way of doing inclusion very different from writing *about* people.

This collection focuses on inclusion in decision-making and social movements, and the authors in several instances utilize case studies to explore forms of organizing inclusive decision-making. They examine participatory models that do not recreate modes of social exclusion. Given the challenges of using participatory methods, it can be especially difficult to do university-based research in this context. As has been observed before, if the university is to be a platform for this mode of knowledge production, the academy must transform in order to value engaged and collaborative methods of research and to provide adequate support for authentically participatory research (Beland Lindahl & Westholm, 2014). Universities certainly need to do work to be more inclusive in their recruiting, hiring, and retention practices, so that university research can be opened up to more viewpoints. Universities will also need to assign greater value to collaborative projects. Paying attention to social exclusions in the research process, especially systematic exclusions, is a value that is in line with the ethos of engaged scholarship. And in order to be sustainable, urban spaces cannot be exclusionary. The goals of urban sustainable development and engaged scholarship are therefore aligned on this point: the production of knowledge about sustainable cities needs to aim at substantive inclusion. Whether the transdisciplinary methods are co-produced, participatory, or simply sensitive to systemic exclusions, their orientation in the urban context should be substantively inclusive.

5. Using Accessible Language

Transdisciplinary research is often expressed in accessible language. This is more of an advantageous by-product of some of the methods employed in TD approaches than a necessary characteristic. Many writers who focus on the democratic importance of TD approaches assume that a high degree of integration between non-specialists and specialists makes knowledge more accessible (Bunders et al., 2015). Therefore, accessibility can result from collaborating with persons from different disciplines or engaging in participatory research. But understanding how to make knowledge accessible can also help us understand what collaboration means. Accessibility can mean creating accessibility for a general or "lay" audience or, alternatively, creating a common language between disciplines. Each kind of accessibility can reveal distinctive features of collaboration.

For example, consciously and reflexively writing a piece for a multidisciplinary audience is one way of creating accessibility. A collection that speaks from different disciplines runs the risk that the same terms can have different meanings at different times, depending on the author's disciplinary background. Even fundamental concepts such as sustainability can be used very differently, as noted earlier. When multiple authors from different disciplines write together, some of the differences of usage may be worked out in the writing so that one definition is offered and used consistently throughout a text. This is not the only way to do it. It is also possible to use definitions in order to signal that a term is not settled and to mark points of disciplinary difference. Definitions also make a text accessible to someone outside the author's field. In this way, differences can be used as a starting point to interrogate what we have unconsciously accepted from our training and to see other ways of approaching complex problems. This way of creating accessibility involves a form of collaboration that differs from strategies that make knowledge accessible by doing away with differences and that involve more ordered and stable collaborative frameworks.

The process of making knowledge accessible can include engagement with editors and readers who do not share a disciplinary bibliography or lexicon. The collaborators in a project can be defined broadly. Instead of understanding knowledge production to end when an article or book is written or even published, we should consider the reader to be part of the process of knowledge production, and therefore a potential collaborator (Benjamin, 1934). The current requirement that scholars produce plans for "knowledge mobilization" exemplifies the multi-step process in producing knowledge that extends beyond the moment of writing a text.

Afterword: Thinking Through Transdisciplinarity in Urban Sustainability

Engagement with a critic or even an interlocutor online is a new form of collaboration, a new way of making research accessible and comprehensible, and perhaps a novel form of transdisciplinarity.

Just as there are many ways to do TD research, there are many ways of expressing knowledge in accessible ways. Jacobs, for example, was quintessentially accessible without following a script for making her writing accessible. In contrast, a transdisciplinary author may render a text accessible by deciding to explain a niche term or, more dramatically, may adopt a theoretical canon different from the one she trained in because it is more accessible. Non-specialist publications are another way of making knowledge accessible, as many of them require the use of simple language. Arendt's articles on the Eichmann trial were originally written for a general, albeit well-educated audience, as they were written for the *New Yorker*, and they remain her most accessible texts. Saskia Sassen frequently publishes more accessible versions of her arguments in online blogs, newspapers, and non-specialist journals.

The concept of accessibility is not entirely straightforward, and the line between simplification and obfuscation is not always clear. A type of shorthand evolves within disciplines when people share the same bibliographies and references; as a result, a niche language tends to develop that can be far more precise than general language. It is simply not always possible to unpack decades of literature on a topic in a short paper, written in general language. Moreover, the use of simple language can obfuscate an argument. Hannah Arendt seems to be an exemplary case in point. There are instances in which simple language has made her meaning more confusing. Her observations, which she called "configurations" in her writing, or "thought trains," are not easily understood. Seyla Benhabib distils the "key terms" of Arendt's method as "configuration" and "crystallization of elements," but these simple language constructions are not necessarily any clearer to a general audience or even to a specialized audience without access to a book by Benhabib that explains what Arendt meant (Benhabib, 2003, p. 64). Making language accessible may involve going outside disciplinary conventions. However, at other times accessibility may require working within conventions. Disciplinary conventions can create epistemic stability in ways that simple, "undisciplined" language may not.

Using accessible language is a political choice that can have important implications for social movements and for democratic processes. A transdisciplinary approach may have the advantage of making conversations more accessible. That is not to say that transdisciplinarity is anti-theoretical or that its insights must always be communicated in simple or generally accessible terminology; rather, transdisciplinary researchers using accessible language can aim at a general audience, can (at times) facilitate collaborative work from different perspectives, and can force the articulation of assumptions.

Conclusion

Cross-disciplinary work is familiar if not basic for urban studies scholars, many of whom sit in schools or institutes that are themselves difficult to categorize in disciplines. They already conduct their research by rejecting disciplinary constraints; they resist the fetishization of expertise; they decompartmentalize systems of knowledge; and they include the excluded. In disciplines outside these special centres, however, research about the sustainability (and security as a subcategory of sustainability) of cities tends to drift towards specialization, even when it is done through a form of interdisciplinarity that relies on the expertise of two or more disciplines. This is a tendency in law. The result is that expert knowledge and professional specialization dominate law's disciplinary conventions, and these conventions in turn channel political choices. I suggest that we should take transdisciplinarity as a methodological choice that pushes back against forces that channel and silo knowledge, potentially frustrating the kinds of political transformations we need.

Returning now to the disciplinary problem with which I set up this methodological reflection, I want to suggest that there are good reasons for urban studies and local government law to occasionally sidestep certain expert vocabularies. One can resist the pull of certain strands of law and economics in this particular moment, for example, for reasons that have nothing to do with being unable to keep current with the literature. If one approaches crime and urban security as an issue of urban sustainability writ large, then it is possible to resist turning this complex social problem into a single-vocabulary or single-logic problem. In contrast, when approaches that fetishize disciplinary logics turn to complex phenomena such as crime, their explanations tend to be flat and single-dimensional, such as rational-choice economic explanations. To take one example, Glaeser and Sacerdote's explanations for the higher rates of crime in high-density urban areas are framed in terms of two reductive alternatives. According to the first rational incentives explanation, economies of agglomeration increase opportunities to financially benefit from committing crime, while lowering the chances of being caught. A second explanation, which rests on a statistical correlation between rates of crime and the number of female-headed households in urban areas, claims that higher rates are explained "by the presence of more female-headed households in cities" (Glaeser & Sacerdote, 1999, p. 253).

While Schleicher (2016) suggests that local government law needs to take on the insights of agglomeration economics in order to be current with developments in scholarship, I suggest that we will be far better prepared to deal with the coming crises if we don't pretend that disciplinary conventions necessitate adopting any particular model or vocabulary for analysis.[4] As the climate crisis exacerbates already existing economic and social inequalities and urban violence, we will urgently need social explanations that take into account multiple factors and the different realities of people's material lives. Inclusivity is the comparative advantage of our approach. Transdisciplinary methods are not a failsafe against simplistic social explanations, but a TD approach – understood as a sensibility rather than rigid method – provides for nuanced and inclusive social explanations. It thus opens the way for political solutions that are better than those that law and economics alone, or its kindred disciplines can provide.

Although transdisciplinarity, as such, has a relatively short history, the move away from intellectual and political silos can be thought of as a return to older forms of knowledge production that were not so rigid, differentiated, or specialized. This collection is the culmination of conversations that started in workshops and conferences among the authors, many of whom are members of the McGill Centre for Interdisciplinary Research on Montreal. The chapters in this book should be read as beginning a conversation among the authors and with the reader, in the spirit of problem-solving. The first step for a transdisciplinary approach is to get a conversation started, and that has happened. We hope that this conversation will continue. Just as researchers are asked to have a particular sensibility when conducting TD research, the reader is also implicated in the research process. Each is asked to read with a sensibility that celebrates experimentation and openness and to become a collaborator in knowledge production. While it runs the risk of expanding the profile of transdisciplinarity too much, this afterword is a celebration of the valuable characteristics of transdisciplinarity as well as a challenge to the tendency in some applications of the approach towards being overly specific. While transdisciplinarity does not always reject disciplinary conventions, it always celebrates the benefits of methodological openness, of challenging disciplinary conventions, of discipline hopping, and of reading across and sometimes against one's own discipline. There is an exciting surge of openness in the academy to discipline-bending as a way of producing knowledge about the serious, complex problems we confront. This spirit

---

4 I could make the same argument Schleicher makes – that his failure to engage seriously with Thomas Picketty's work is a failure to keep up with the current literature. It is, however, clear that Schleicher only has passing references to Picketty's work because he doesn't find Picketty as convincing as he does Glaeser. Choices of bibliographies often only look like disciplinary necessity when they are politically mainstream or support the status quo.

of openness, flexibility, reflexivity, and inclusion should open new ways of thinking about the environmental and social problems we face, and with any luck, it will open new ways of acting in the face of crisis.

REFERENCES

Agarwal, B. (2001). "Participatory exclusions, community forestry, and gender: An analysis for South Asia and a conceptual framework." *World Development, 29*(10), 1623–1648. https://doi.org/10.1016/S0305-750X(01)00066-3

Agger, A. (2012). Towards tailor-made participation: How to involve different types of citizens in participatory governance. *Town Planning Review, 83*(1): 29–45. http://online.liverpooluniversitypress.co.uk/doi/10.3828/tpr.2012.2

Arendt, H. (1951). *The origins of totalitarianism.* Harcourt Brace.

Arendt, H. (1994). We refugees. In M. Robinson (Ed.). *Altogether elsewhere: Writers on exile* (pp. 110–119). Faber and Faber.

Beland Lindahl, K., & Westholm, E. (2014). Transdisciplinarity in practice: Aims, collaboration and integration in a Swedish research programme. *Journal of Integrative Environmental Sciences, 11,* 155–171. https://doi.org/10.1080/1943815X.2014.945940

Benhabib, S. (2003). *The reluctant modernism of Hannah Arendt.* Rowman & Littlefield New.

Benjamin, W. (1934). The author as producer. Address at the Institute for the Study of Fascism.

Bunders, J. F. G., Bunders, A. E., & Zweekhorst, M. B. M. (2015). Challenges for transdisciplinary research. In B. Werlen (Ed.), *Global sustainability: Cultural perspectives and challenges for transdisciplinary integrated research* (pp. 1–300). Springer International.

Buxton, G. V. (1992). Sustainable development and the summit: A Canadian perspective on progress. *International Journal, 47*(4), 776–795. https://doi.org/10.2307/40202803

Canada, Environment and Climate Change. (2019). *Canada's changing climate report.*

CBC News. (2019, 23 September). Plante tells United Nations that Montreal will reduce emissions by 55% by 2030. https://www.cbc.ca/news/canada/montreal/valerie-plante-united-nations-address-climate-change-1.5293829

Ellis, G., Monaghan, J., & McDonald, L. (2015). Listening to "generation Jacobs": A case study in participatory engagement for a child-friendly city. *Children, Youth and Environments, 25*(2), 107–127. https://doi.org/10.7721/chilyoutenvi.25.2.0107

Friedmann, J. (1970). Review: The economy of cities by Jane Jacobs (Random House, 1969). *Urban Affairs Quarterly.*

Fritz, L. (2016). (De-)constructing participation in transdisciplinary sustainability research: A critical review of key concepts." In J. Engelschalt, A. Maibaum, F. Engels, & J. Odenwald (Eds.), *Schafft Wissen: Gemeinsames Und Geteiltes Wissen in Wissenschaft* (pp. 106–124). SSOAR Open Access Repository.

Glaeser, E. L., & Sacerdote, B. (1999). Why is there more crime in cities? *Journal of Political Economy, 107*, 225–258. http://doi.org/10.1086/250109

Goldberger, P. (2006). Urban planning: Uncommon sense: Remembering Jane Jacobs, the 20th century's most influential critic. *The American Scholar, 75*(4), 122–126. https://www.jstor.org/stable/41222661

Hirsch Hadorn, G., Hoffmann-Riem, H., Biber-Klemm, S., Grossenbacher-Mansuy, W., Joye, D., Pohl, C., Wiesmann, U., & Zemp, E. (2008). The emergence of transdisciplinarity as a form of research. In *Handbook of transdisciplinary research* (pp. 19–39). Springer.

Intergovernmental Panel on Climate Change. (2018). *Global warming of 1.5°C*. https://www.ipcc.ch/sr15.

Jacobs, J. (1961). *The death and life of great American cities*. Vintage.

Kennedy, D. (2018). *A world of struggle: How power, law, and expertise shape global political economy*. Princeton University Press.

Kiteme, B. P., & Wiesmann, U. (2008). Sustainable river basin management in Kenya: Balancing needs and requirements. In G. Hirsch Hadorn, H. Hoffmann-Riem, S. Biber-Klemm, W. Grossenbacher-Mansuy, D. Joye, C. Pohl, U. Wiesmann, & E. Zemp (Eds.), *Handbook of transdisciplinary research* (pp. 63–78). Springer.

Krieger, L. (1976). The historical Hannah Arendt. *The Journal of Modern History, 48*(4), 672–684. https://www.jstor.org/stable/1880198

Laurence, P. L. (2007). Jane Jacobs before *Death and Life*. *Journal of the Society of Architectural Historians, 66*(1), 5–15. http://doi.org/10.1525/jsah.2007.66.1.5

Lilja, N., & Bellon, M. (2008). Some common questions about participatory research: A review of the literature. *Development in Practice, 18*(4), 479–488. http://doi.org/10.1080/09614520802181210

Macdonald, R. (2000). Perspectives from legal theorists. In M. A. Somerville & D. J. Rapport (Eds.), *Transdisciplinarity* (pp. 61–107). McGill-Queen's University Press.

Mobjörk, M. (2010). Consulting versus participatory transdisciplinarity: A refined classification of transdisciplinary research. *Futures, 42*(8), 866–873. http://doi.org/10.1016%2Fj.futures.2010.03.003.

Møhring Reestorff, C., Fabian, L., Fritsch, J., Stage, C., & Løhmann Stephensen, J. (2014). Introducing cultural participation as a transdisciplinary project. *Conjunctions, 1*(1): 1–25. http://doi.org/10.7146/tjcp.v1i1.18601

Newig, J., Derwort, P., & Jager, N. W. (2019). Sustainability through institutional failure and decline? Archetypes of productive pathways. *Ecology and Society, 24*(1). http://doi.org/10.5751/ES-10700-240118

Nickles, T. (2003). Evolutionary models of innovation and the Meno problem. In L. V. Shavinina (Ed.), *The international handbook on innovation* (pp. 54–78). Elsevier Science.

Osborne, P. (2015). Problematizing disciplinarity, transdisciplinary problematics. *Theory, Culture & Society, 32*(6), 3–35. https://doi.org/10.1177%2F0263276415592245

Pennington, D. D., Simpson, G., McConnell, M., Fair, J., & Baker, R. (2013). Transdisciplinary research, transformative learning, and transformative science. *BioScience, 63*(7), 564–573. http://doi.org/10.1525/bio.2013.63.7.9

Pohls, C., & Hirsch Hadorn, G. (2007). *Principles for designing transdisciplinary research.* Oekom. https://doi.org/10.14512/9783962388638

Polk, M., & Knutsson, P. (2008). Participation, value rationality and mutual learning in transdisciplinary knowledge production for sustainable development. *Environmental Education Research, 14*(6), 643–653. http://www.tandfonline.com/doi/abs/10.1080/13504620802464841

Reestorff, C. M., Fabian, L., Fritsch, J., Stage, C., & Stephensen, J. L. (2021). Conjunctions: Introducing cultural participation as a transdisciplinary project. *Sciendo, 1*(1), 1–25. https://www.sciendo.com/article/10.7146/tjcp.v1i1.18601

Sassen, S. (2013a). Before method: Analytic tactics to decipher the global – An argument and its responses, part I. *The Pluralist, 8,* 101–112. https://www.jstor.org/stable/10.5406/pluralist.8.3.0079

Sassen, S. (2013b). Before method: Analytic tactics to decipher the global – An argument and its responses, part II. *The Pluralist, 8,* 101–112. https://doi.org/10.5406/pluralist.8.3.0101

Sassen, S. (2014). *Expulsions: Brutality and complexity in the global economy.* Harvard University Press.

Schleicher, D. (2016). Local government law's "law and ———" problem. *Fordham Urban Law Journal, 40*(5), 1951–1973.

Simon, D., Palmer, H., Riise, J., Smit, W., & Valencia, S. (2018). The challenges of transdisciplinary knowledge production: From unilocal to comparative research. *Environment and Urbanization, 30*(2), 481–500. https://doi.org/10.1177%2F0956247818787177

Simpson, L. (2008). Looking after Gdoo-naaganinaa: Precolonial Nishnaabeg diplomatic and treaty relationships. *Wicazo Sa Review, 23*(2), 29–42. http://doi.org/10.1353/wic.0.0001

Van Duzer, C. H. (1952). Reviewed works: The origins of totalitarianism by Hannah Arendt. *The American Historical Review, 57*(4), 933–935. https://doi.org/10.1086/ahr/57.4.933

Voegelin, E. (1953). The origins of totalitarianism. *The Review of Politics, 15*(1): 68–76.

Contributors

Bria Aird is a professional planner (RPP MCIP) at Fotenn Planning + Design and is based in Ottawa, Ontario. Her practice includes shopping mall redevelopment, mixed-use intensification, and small-scale infill. Outside her nine-to-five, Bria volunteers with civil society groups, including acting as an expert advisor for urbanism-related student research through the Ottawa Eco-Talent Network.

Deborah Curran is an associate professor at the University of Victoria in the Faculty of Law and the School of Environmental Studies. Deborah teaches courses related to land and water, including municipal and water law, and her research focuses on adapting water law and sustainable land use. As the executive director with the Environmental Law Centre at UVic, Deborah supervises students working on projects for community and Indigenous organizations across the province. For over twenty years she has worked with local governments and community organizations on creating sustainable communities through the implementation of green bylaws. Deborah is the author of the Green Bylaws Toolkit, and she co-founded Smart Growth BC.

Natalya Berezina Dreszer graduated with a B.A.Sc. in environment from McGill University, where she worked in Kevin Manaugh's lab. She now works in local journalism, focusing on centring the needs of local communities in government reporting. She lives with her family in Santa Cruz, California.

Jonathan Durand Folco is an assistant professor at the Élisabeth-Bruyère School of Social Innovation at Saint Paul University, Ottawa. His research interests include participatory democracy, municipal politics, the commons, and socio-ecological transition. He is the author of the book *À nous la ville! Traité de municipalisme* (Écosociété, 2017), co-author of *Manuel pour changer le monde* (Lux, 2020), and editor of *Montréal en chantier: les défis d'une métropole pour le XXIe siècle* (Écosociété, 2021).

Alexandra Flynn is an assistant professor at the University of British Columbia's Allard School of Law. Her teaching and research focus on municipal law and governance, and property and administrative law. She has published numerous peer-reviewed papers and reports on how cities are legally understood in Canadian law and how they govern, covering topics such as the constitutional role of municipalities and the legal relationships among First Nations and municipal governments. She is currently working on the legal rights of precariously housed people in Canadian cities.

Hoi L. Kong is the inaugural holder of the Rt. Hon. Beverley McLachlin, P.C., UBC Professorship in Constitutional Law, which he assumed in 2018. He researches and teaches in the areas of constitutional, administrative, municipal, and comparative law, and constitutional and public law theory. Prior to joining the Allard School of Law, Kong was an assistant and then associate professor at McGill University's Faculty of Law, where he served a term as associate dean (academic). He was previously an assistant professor of law, cross-appointed with the School of Urban Planning at Queen's University, and an associate-in-law at the Columbia Law School. Kong clerked for Justice L'Heureux-Dubé and Justice Deschamps at the Supreme Court of Canada.

Nina-Marie Lister is a professor and graduate director in the School of Urban and Regional Planning at Toronto Metropolitan University, where she founded and directs the Ecological Design Lab. A senior fellow of Massey College, she is also visiting professor of landscape architecture at Harvard University. She holds the Margolese National Design for Living Prize for her work in ecology and design, and she was awarded honorary membership in the American Society of Landscape Architects. Lister is co-editor of *The Ecosystem Approach* (Columbia, 2008) and *Projective Ecologies* (Actar, 2014, 2020) and author of more than one hundred scholarly and professional publications. Her work connects people to nature in cities, through green infrastructure design for climate resilience, biodiversity, and human well-being.

Nik Luka is cross-appointed to the Peter Guo-hua Fu School of Architecture and the School of Urban Planning at McGill University, where he is also the associate director of the Centre for Interdisciplinary Research on Montréal. As an ethnographer specialized in housing, social practice, and landscape studies, he works in close collaboration with civil-society organizations and state agencies on co-production through community-based design.

Kevin Manaugh is an associate professor jointly appointed in the Department of Geography and the Bieler School of Environment at McGill University. His research focuses on the equity and justice dimensions of transport infrastructure and policy.

Tanya Monforte is an assistant professor in international law and society at Concordia University in Montréal. The former director of the human rights program at the American University in Cairo, she has held visiting positions at universities around the world. She writes and publishes on human rights, security, and legal theory.

Richard Shearmur is an economic geographer and urban planner whose research focuses on development processes in peripheral regions, on trying to locate mobile jobs within metropolitan areas, and on the connection between innovation and local development. It is in this latter context that non-market innovation within and by municipalities has become a focus of his work. He enjoys food, and cycles in order to eat. He is also a professor (since 1999) and is currently director of the McGill School of Urban Planning.

Index

Figures and tables are indicated by page numbers in italics.

Aalborg Charter, 21–2
access and accessibility: and BIAs, 145; and bike lanes, 154; and charrettes, 173, 174; and city governance, 3; and collaborative governance, 126; and commons, urban, 121; and consultation processes, 132–3; and deliberative democracy, 110, 134; and innovation, 66, 67, 81; and pedestrian spaces, 7, 44, 46, 47, 51, 52, 55, 59–60; and smart cities, *16*, 20, 23, 28; and transdisciplinarity, 200, 206, 210–12
accountability: and collaborative governance, 126, 127; and smart cities, 19, 31; and stakeholder groups, 138, 145–6, 148, 155, 157; and sustainable development, 23, 132, 134; and Turcot Interchange, 52, 59. *See also* care, duty of
activism. *See* social movements
ad hoc decision-making, 9, 138, 144, 156, 183
affordable housing, 150
agency, 47, 120, 143, 163n2, 166n8, 167, 170n14, 177. *See also* autonomy, individual
agglomeration economics, 196, 212

agreement, 5, 6, 72, 82, 122, 125, 127, 182. *See also* collective action problems
Agricultural Land Reserve (ALR), 97
Agyeman, Julian, 143, 169, 173
Airbnb, 18, 19, 31, 32, 76
Aird, Bria, 9
air quality, 44, 123, 124, 131, 153
Alberta, 85
Alexander, Gregory, 141
alleys, 47, 55–8, 59, 84–5
Alliance pour un nouveau partage de la rue Sainte-Catherine, 60
allied design professions, 163. *See also* experts or professionals
Alphabet, 18
Amin, Ash, 13
Annex Business Bike Alliance (ABBA), 155, 156
Annex Residents Association, 154
Applebaum, Michael, 52
ArcGis software, 83
architects and architecture: and charrettes, 9, 163, 165, 167, 168, 171, 176; and green alleys, 56; and Mirvish Village, 149; and urban studies, 4. *See also* Mirvish redevelopment; Turcot Interchange

Arendt, Hannah, 199–200, 201–2, 203, 205, 208–9, 211
Arnstein, Sherry R., 20–1, 25, 33, 46, 165
artificial intelligence (AI), 20, 28. *See also* digital technologies
arts and culture, 54, 148, 151–2
assemblage theory, 172n18
associations. *See* neighbourhood associations
austerity urbanism, 19–20
automobiles. *See* vehicles
autonomy, individual, 21, 25, *34*, 95–6. *See also* agency

Bacqué, Marie-Hélène, 33
Barcelona, 30, 31, 32
Basera, Neelam, 21
Benhabib, Seyla, 201–2, 211
Berezina Dreszer, Natalya, 7
Bernstein, Scott, 123
Berry, Jeffrey, 139
Berthierville (QC), 80
best practices, 31, 97, 111, 171–2, 197
bibliographies, 196, 197, 202, 205, 210, 211, 213n4
Bibri, Simon Elias, 17
big data, 13, 15, 17, 19, 77. *See also* data
bike lanes, 138, 144, 153–7. *See also* cycling
biodiversity, 22, 55, 107, 108–9, 111, 113, 194
bio-methanization, 80
Blair, Tony, 28
Blais, Pamela, 97–8
Blanchet-Cohen, Natasha, 182
Blank, Yishai, 3
Blé (local currency), 32
Bloor Annex BIA (Toronto), 151, 153, 155
Bloor St. Culture Corridor, 154

Bloor Street West (Toronto), 137, 138, 149, 153–7
Bologna, 122, 125
borders and boundaries: and charrettes, 163, 172, 175; and DPAs, 103; and knowledge, 205; and local governance, 138, 139, 141; and municipalities, 69, 118n1; and private property, 112; and stakeholder groups, 145, 146, 147; and transdisciplinarity, 196–7, 198–200; and Turcot Interchange, 48; and urban growth, 96, 97, 104, 111
borough governance: definitions and features, 68–70; and green alleys, 56–8; and innovation, 65–6, 67–8, 74, 78–9, 84–5, 86–7; in Montreal, 49–50, 54; and Turcot Interchange, 51, 52, 53, 59. *See also* local governance
Bos, Nathan, 167
Bourdieu, Pierre, 171
brainstorming, 175, 177, 178, 180
Bria, Francesca, 16, 19, 20, 31, 33, 34
Brissenden, Annemarie, 149, 151
British Columbia (BC), 95, 96, 97–9, 102–4, 110, 112
Brown, Wendy, 19
Brundtland Commission, 97
Buchanan, Richard, 163n1
budgets and budgeting, 24, 33, 74, 75, 83, 84, 85, 146
Building Sustainable Communities (Roseland), 97
Bunders, Joske, 197, 204
Burch, Sarah, 178
Burchell, Brian, 151
Burgen, Stephen, 31
Burke, Edmund, 205
Burner Alley, 57–8
burst effect, 167, 168, 171, 174, 183. *See also* charrettes, transdisciplinary

business-as-usual, 97, 171, 172, 174, 177
business improvement areas (BIAs), 140, 141, 145, 146–8, 151–3, 154–7. *See also* borough governance; stakeholder groups

Canada: and cities, definition of, 118n1; and climate change, 194; and extractive technologies, 86; municipalities in, 65, 68–9; property rights in, 8, 97–101, 102, 104, 112; and rebel cities, 31; roads and transportation in, 44–5, 46; and smart cities, 24; snow removal in, 83; sustainable development in, 4, 5, 157; zoning in, 128
Canada Lands Inventory, 97. *See also* land use and regulation
capitalism, 17, 22, 28, 30, 32, 35, 71n4, 87
carbon neutrality, 98–9. *See also* greenhouse gases
Cardullo, Paolo, 20, 21, 25, 33
care, duty of, 72, 76, 83, 86. *See also* accountability
cars. *See* vehicles
Carter, Adam, 18n1
Centre for Interdisciplinary Research on Montreal, 213
Chapman, Matthew, 57
charrettes, transdisciplinary: and citizen participation, 165–6; and co-production, 166–7, 170–4; definitions and context, 9, 140, 167–70; examples of, 182–3; and inclusion, 208; key advantages of, 174–8; overview, 163–5; procedural tips for, 178–81. *See also* transdisciplinarity
Charter of European Sustainable Cities, 21–2

Chaskin, Robert, 141
Chemerinsky, Erwin, 139–40
Chicago, 141
CITIES, 31. *See also* rebel cities
cities, legal definition, 118n1. *See also* commons, city as; rebel cities; smart cities; *specific cities*
citizen participation, definitions, 14, 195. *See also* borough governance; collaboration and collaborative governance; consultations; innovation; local governance; smart cities; stakeholder groups
Citizens for Responsible EDPA, 105–6
city governance. *See* borough governance; collaboration and collaborative governance; local governance
Climate Action Charter (BC), 95, 98–9
climate change, 33, 87–8, 97–8, 111, 178, 194, 213. *See also* environmental challenges
Climate Change Accountability Act, 98
Clinton, Bill, 28
coffeehouse culture, 133
Cohen, Boyd, 17–18
collaboration and collaborative governance: and accessible language, 210–12; and charrettes, 164–5, 166, 169, 170, 171, 173, 176, 178, 180, 182, 183; and commons, urban, 121–2, 124, 125–6, 133–4; deliberative democratic criticism of, 126–8, 133–4; and innovation, 66, 71, 119; and knowledge production, 209; in Montreal, 59, 60, 131–3; and rebel cities, 30, 31, 33; and smart cities, 7, 13–14, 20, 21, 23–6, 27–9, 35–6; and transdisciplinarity, 9, 207, 208, 213; and wicked problems, 163n2, 164. *See also* co-production; stakeholder groups

collective action problems: and charrettes, 163, 166–7, 173, 177–8; and climate change, 194–5; and commons, urban, 121–2; and innovation, 82, 84, 85, 87–8; and local governance, 139, 143–4; and municipalities, 70; and pedestrian spaces, 59; and property rights, 95–6, 104–12, 113; and stakeholder groups, 125, 141, 148, 155–6; and sustainable development, 5–6, 8–9; and transdisciplinarity, 196, 204; and transportation, 49–50. *See also* law and policy

commons, city as, 31–2, 118, 119–22, 122n4, 124–6, 130–2, 133–4

community, meaning of, 141. *See also* local governance

commuting, 123, 124, 130–1. *See also* transportation

compensation, 8, 100–1, 104. *See also* property rights

competition (for resources), 18, 19, 29, 66, 67, 70

competitions (organized events), 68, 74n6, 79, 87, 168n11, 170, 176, 180

complexity, brutality of, 205. *See also* transdisciplinarity

compression effect, 9, 174, 178, 179, 183. *See also* charrettes, transdisciplinary

computer software, 83–4. *See also* digital technologies

Concorde Overpass, 48

Conference on Climate Change (COP 21), 128–9

constructivism, 176

consultations: and commons, urban, 122, 130–1; and innovation, 75; and knowledge production, 207; legal context of, 128;

meanings of, 195; in Montreal, 8, 50, 53, 54, 59, 118–20, 128–33; and municipalities, 69, 70; in Saanich, 106, 108; and smart cities, 20, 25, *26*, *36*; and stakeholder groups, 137, 147–8; in Toronto, 149–55, 156–7. *See also* collaboration and collaborative governance

consumption and consumerism: and commons, urban, 121, 130–1; and innovation, 66–7; and rebel cities, 33; and smart cities, 20–1, *22*, *26*, *36*; and software innovation, 83; and sustainable development, 8, 23, 25, 27, 99; and urban development, 123, 131

cooperative housing, 182n23

co-production: and charrettes, 164, 167, 170–4, 176, 180, 183; and citizen participation, 166; definition, 164n3; and inclusion, 209; and transdisciplinarity, 207. *See also* collaboration and collaborative governance; stakeholder groups

Corner, James, 163

COVID-19 pandemic, 183

creative destruction, 71. *See also* innovation

crime, 212

cross-fertilization, 81. *See also* innovation

Croteau, François, 84

crowdsourcing, 31, 166, 174, 182, 183. *See also* collaboration and collaborative governance; co-production

cultural competency, 173. *See also* charrettes, transdisciplinary

Curran, Deborah, 8, 206

currencies, 32

Cycle Toronto, 153
cycling: in Canada, 45; and GAHN project, 182; in Montreal, 47, 54; and stakeholder groups, 144; and sustainable development, 143; in Toronto, 138, 144, 152, 153–7

Dale, Ann, 4, 166n8, 167
Dalle-Parc, 47, *49*, 51, 52–3, 59
Darchen, Sébastien, 148
Dardot, Pierre, 18
data: and consultation, 133; and innovation, 77; and rebel cities, 32, 33, 34; and smart cities, 15, *16*, 17–18, 19–20, 21, *22*, *36*; and stakeholder groups, 140, 146; and sustainable development, 13, 21, 23–4, 31, 138
Davidoff, Paul, 165
Daviter, Falk, 163n1
Death and Life of Great American Cities, The (Jacobs), 201
Decide Madrid, 33
decision making: in Canada vs. France, 68–9; challenges of, 5, 9; and charrettes, 163, 165, 170, 172–3, 177, 178, 181; and citizen participation, 47, 56, 58, 107, 109, 110, 129; and collaborative governance, 25, 28, 122, 126–8, 133; and commons, urban, 121, 122; and inclusion, 46, 209; and innovation, 86; in Montreal, 50, 164n3; Nishnaabeg, 198–9n2; and political support, 59; and rebel cities, 32, *34*, *36*; and smart cities, 20, 21; and stakeholder groups, 138–9, 141, 142, 143, 144–5, 147–8, 155, 157; and sustainability, 22, 78, 166n8; and sustainable development, 98, 123, 125; and transdisciplinarity, 171n17, 204

decompartmentalization, 200, 204–6, 212. *See also* transdisciplinarity
definitions, disciplinary, 210. *See also* transdisciplinarity
degrowth, 195. *See also* economic growth
de la Concorde Overpass, 48
Deleuze, Gilles, 167
deliberative democratic theory, 8, 110, 119–20, 126–8, 132–4
democratic processes: and charrettes, 167, 172, 173, 182; and collaborative governance, 8, 119, 126–8, 132–4; and experimentalism, 118–19; legal theory of, 120; and neoliberalism, 18–19; and property rights, 110; and rebel cities, 29–30, 31, 32, 33–4, *36*; and smart cities, 15, 20, 25, 26, 27, 29, 35; and social movements, 194–5; and stakeholder groups, 139, 140, 141, 142, 146; and transdisciplinarity, 204, 210, 211
density and densification: in British Columbia, 96, 98, 99, 100, 103–4, 111, 112; and crime, rates of, 212; and smart cities, 15; in Toronto, 150, 152, 153, 157; and urban sprawl, 123. *See also* growth management
Dernbach, John C., 123
design, definition of, 172
design development, 165, 167, 168, 170–1, 173, 174–5, 180
design thinking, 119
development permit areas (DPAs), 96, 101–4. *See also* property rights
Dexter, Sue, 152
dialogue, 4, 69, 175, 179, 181
Diamond Head Consulting, 108
digital technologies: and capitalism, 28; and commons, urban, 122; and innovation, 66, 77, 83–4;

digital technologies (*continued*)
  and rebel cities, 31–2, 33–5; and smart cities, 13–16, 17–18, 19–20, 21, 35–6; and sustainable cities, 22, 23–4, 25–6, 27
disciplines. *See* transdisciplinarity
discourse analysis, 14
Dockside Green (Victoria), 99, 100
Downtown Yonge BIA (Toronto), 140
Drapeau, Jean, 55
Drummondville (QC), 80
Dublin, 20, 25
Durand Folco, Jonathan, 6–7
duty of care, 72, 76, 83, 86. *See also* accountability

EcoDensity initiative, 98, 99, 111. *See also* density and densification
École des Beaux-Arts, 169
economic growth: and innovation, 66–7, 70–1; and neoliberalism, 18; and rebel cities, 33; and smart cities, 14, 21, 28–9; and sustainability, 4, 78, 148, 195; and sustainable cities, 23, 27, 31
Economy of Cities, The (Jacobs), 203
education: and charrettes, 140, 169, 176, 180; and citizen participation, 125; and collaborative governance, 132; and innovation, 80; and pedestrian spaces, 46; and smart cities, *16*, 24; and sustainability, 21
empathy, 110, 172, 180
empowerment: and charrettes, 167, 170, 177; and innovation, 68, 84, 87; and local governance, 54; and rebel cities, 33; and smart cities, 6, 7, 14, 20, 23, 24–5, 28, *36*; and stakeholder groups, 144. *See also* power relations
engagement. *See* collaboration and collaborative governance; consultations; co-production; stakeholder groups
Engle, Jayne, 182, 183
Enlightenment, 172
Entertainment District BIA (Toronto), 148
entrepreneurship: and collaborative governance, 122n5, 125; and innovation, 66, 67, 71–2, 73–5, 81; and neoliberalism, 19; and smart cities, 14, 15–16, 20, *36*; and sustainable cities, 24, 28
Environmental Assessment Board (BAPE), 50, 51, 52, 59
environmental challenges: and automobiles, 44; and charrettes, 163, 178; and governance, 3; and innovation, 87–8; and land-use laws, 123; and property rights, 95–6, 110–12; and rebel cities, 33; and smart cities, 14–15, 17, 35–6; and sustainability, 78; and sustainable cities, 21, 22–3, 27; and sustainable development, 4, 97–8, 102, 107, 130–1; and transdisciplinarity, 194–5, 196–7, 199, 205, 213–14. *See also* greenhouse gases
environmental development permit areas (EDPAs): and citizen participation, 109–12; definitions, 97, 101–2; and property rights, 102–4, 105–9; in Saanich, 96, 104–5, 113
environmental studies, 4, 197, 198
equilibrium, 21, 26–7. *See also* sustainability
equity: and commons, urban, 124; and innovation, 85; and social inclusion, 46; and sustainability, 4, 78, 143; and sustainable development, 22, 23, 27, 29, 79
Erickson, Jennifer S., 173

Espiau, Gorka, 119
Esplaneuve Green Alley, 57. *See also* green alleys
ethics, 18, 34, 46, 72, 78, 138, 198–9n2
European Smart Cities, 15. *See also* smart cities
Evans, Tom, 143
evolutionary economics, 71n5. *See also* economic growth
experimentalism, 118–19. *See also* innovation
expertise, fetishization of, 202–4
experts or professionals: and charrettes, 9, 163–5, 167–8, 169–71, 173–8, 181–3; citizens as, 147; and consultation processes, 128, 131–2, 147; and land regulation, 104, 108, 109, 110; and smart cities, 23, 25; and transdisciplinarity, 196, 198, 200, 202–5, 207, 210–11, 212
expropriation, 100–1, 103–4. *See also* property rights
Expulsions (Sassen), 205

Fagotto, Elena, 139
Fainstein, Susan, 29
FairBnB, 32
fairness, 107, 108, 109, 110, 112. *See also* property rights
farmland, 97. *See also* land use and regulation
"Fearless Cities" (international summit), 30
Federation of North Toronto Residents Associations, 155
feedback loops, 71, 163, 175, 181
female-headed households, 212
financial crises, 29–30
firms, 69, 71, 73–4, 75–6, 77. *See also* stakeholder groups
First Nations, 6n3, 198–9n2

fish and fish habitats, 102, 103, 104
Fix-Your-Street, 25
Flynn, Alexandra, 3–4, 8–9, 206
Folco, Durand, 202
Ford, Cristie, 118–19
Ford, Richard Thompson, 140
Forester, John, 173, 179, 183
fossil fuels, 8, 118, 122, 128–31. *See also* greenhouse gases
Foster, Sheila, 121–2, 124, 125, 126, 128, 131
Foucault, Michel, 18
framing and reframing, 175. *See also* charrettes, transdisciplinary; innovation
France, 69
Freeman, Gabrielle, 15, 21, 24, 26, 27, 28, 33
free zones, 32. *See also* rebel cities
Freyfogle, Eric T., 112
Friedmann, John, 203–4
Friends of the Earth Europe (FoEE), 32
Fritz, Livia, 207, 208
Frug, Gerald, 140
Fung, Archon, 25, 139
future generations: and commons, urban, 122, 130; and democratic processes, 8; and experts, decisions of, 163–4; and Nishnaabeg ethics, 198–9n2; and sustainability, 78, 122–3; and sustainable development, 3, 6, 21, 23, 24; and transdisciplinarity, 195

Gamon, Julia A., 180
gardens and gardening, 54, 84, 106, 107
Garg, Sunil, 141
gas, methane, 80. *See also* greenhouse gases
gentrification, 70, 137, 151–2

Geographic Information Systems (GIS), 138
Gibson-Graham, J. K., 68, 87
Giddens, Anthony, 28
Glaeser, Edward L., 212
globalization, 19
Godin, Benoît, 75
Goldberger, Paul, 202
Google, 18, 20, 77
governance vs. government, 125. *See also* borough governance; local governance
governmentality, 18. *See also* neoliberalism
GPS, 83
Gravel, Felix, 52
Green, Active, Healthy Neighbourhood (GAHN), 182–3
green alleys, 47, 55–8, 59, 84–5
Greenberg, David, 141
green building, *16*, 99
greenhouse gases (GHG): and bike lanes, 153; and climate change, 194; and collective action, 8; and innovation, 80, 84; and pedestrian spaces, 45; and sustainable development, 95, 97, 98–9, 130–1; and urban development, 123. *See also* environmental challenges
Greenhouse Gas Reduction Targets Act, 98
green roofs, 100, 113
Grenoble (France), 24
grounding, 176, 179
growth management, 15, 96–7, 98–9, 101, 104, 111, 123, 125. *See also* density and densification
Guattari, Félix, 167
Gutmann, Amy, 120n2, 127n10

Haarstad, Håvard, 15–16
Habermas, Jürgen, 133, 139
hackathons, 19, 31, *34*, 130

Hagan, Margaret, 119
Harbord Village Residents' Association (HVRA), 150, 152, 154
Harris, Andrew, 78n8, 87
Harvey, David, 59
Haussmannization, 54
Hawkins, Christopher, 125
highways, 44, 45–6, 50, 79, 97. *See also* Turcot Interchange; vehicles
Hirsch Hadorn, Gertrude, 197
Hollands, Robert, 19
Honest Ed's, 148, 149–50, 154
humility, 180, 183
Hutchinson, Brian, 108

Iaione, Christian, 121–2, 124, 125, 126, 128, 130–1
imitation, 75. *See also* innovation
inclusion and inclusivity: and charrettes, 166, 167; and commons, urban, 124; and consultation processes, 128, 138; and law, 5, 8; in Montreal, 7, 164n3, 182; and pedestrian spaces, 45, 46, 58, 60; and smart cities, *16*, 20, 23, *36*; and sustainable development, 4, 98, 100; and transdisciplinarity, 9, 195, 199, 206–9, 213, 214
India, 15
Indigenous perspectives, 6n3, 198–9n2
information and communications technologies (ICT), 15, 23, 24, 26, 31
infrastructure, urban: and commons, urban, 122; and experts, decisions of, 163–4, 172n18; in Montreal, 7, 45–52, 55, 59, 59–60; and municipalities, 68, 69; and rebel cities, 31–2, 34; and smart cities, 14, 15, 18, 19–20, 27; and

sustainable development, 95, 96, 98, 100, 113, 164–5n5; in Toronto, 150, 153, 155, 157; violence of, 44, 53–4
innovation: and charrettes, 178, 180; and consultation processes, 129–30; definitions and context, 66–7, 75, 118, 195–6, 198–9n2; and entrepreneurs, 73–5; evaluation and politics of, 85–8; examples of, 80–5; and municipalities, 7, 65–6, 67–8, 72–3, 75–7, 85–7, 203; regulation of, 118–19; Schumpeterian, 70–2; and stakeholder groups, 144; and sustainability, 78–9; and sustainable development, 5; and transdisciplinarity, 198–200, 201, 205
intellectual property, 66, 77, 83
interdisciplinarity: and charrettes, 172; and co-production, 166; and law, 196; and local governance, 139; and sustainability, 164; and transdisciplinarity, 179, 201, 212; and urban studies, 4
interest groups. *See* stakeholder groups
International Municipalist Summit, 30
International Telecommunication Union (ITU), 24

Jackson (Mississippi), 30
Jacobs, Jane, 139, 165, 199–200, 201, 202–4, 207, 211
Jacques Cartier Bridge, 47
jaywalking, 45. *See also* walking
Jevons paradox, 27. *See also* consumption and consumerism

Kafka, Franz, 172n18
Kastelle, Tim, 67n2
Katz, Cindi, 172
Katz, Larissa, 112
Kim, Sohee, 140

Kitchin, Rob, 17, 18, 20, 21, 24, 31, 33
Kiteme, Boniface, 208
Kleiner, Sam, 139–40
knowledge, compartmentalization of, 204–6
knowledge production: and charrettes, 172, 175, 176–7; and inclusion, 209; and interdisciplinarity, 201; and readers, 195, 210, 213; and sustainable development, 9; and transdisciplinarity, 179, 196, 197, 199–200, 205, 207–8, 213
Kong, Hoi, 3–4, 8, 206
Korea Town BIA (Toronto), 153, 155
Krieger, Leonard, 203
Kurdistan, Syrian, 30

Lachine Canal, 48, 49
land use and regulation: in BC, 8, 95, 97–9; in Canada, 97, 99–101; and consultation processes, 128; and EDPAs, 96, 101–9, 113; and innovation, 81, 82, 198n2; and municipalities, 69; and pedestrian spaces, 45; and property rights, 8, 97, 99–101, 109–13; and stakeholder groups, 142, 144, 148; and sustainable development, 119–20, 143, 163; and urban commons, 121, 123–5, 130–1
language: and charrettes, 180; and citizen participation, 46, 108; and consultation processes, 128; and smart cities, 19, 20; and transdisciplinarity, 195, 200, 210–11
Laval, Christian, 18
law and policy: and collective action, 8, 167; and commons, urban, 121; and consultation processes, 119–20, 128, 129; and EDPAs, 101–4, 106, 108–9, 110–11, 113;

law and policy (*continued*)
  and green alleys, 56, 84; and innovation, 118–19; and legitimacy, 127; and local governance, 138, 139, 156; and municipalities, 68–9; and pedestrian spaces, 45; and property rights, 95–7, 100–1, 106, 109, 112–13; social context of, 134; and stakeholder groups, 145–6; and sustainability, 122–3; and sustainable development, 3, 4–5, 97–8; and transdisciplinarity, 196, 212; and Turcot Interchange, 52
Layard, Antonia, 138, 156
Leadership in Energy and Environmental Design, 99
Lefevbre, Henri, 44, 46
legal design, 119. *See also* law and policy
legitimacy, 28, 120, 126–7, 128, 141, 166, 178
Lennertz, Bill, 180
Lewis, Nathaniel M., 142
libraries, 68, 69, 79
Lippert, Randy, 142
Lister, Nina-Marie, 9
Little Burgundy (Montreal), 53–4
Little Burgundy Coalition, 54
living labs, 28
Local Agenda, 21
local governance: and collaboration, 8, 122, 125, 127; definitions, 69, 138; and innovation, 67, 87; in Montreal, 50, 53, 182; and municipalism, 30; and property rights, 101–4, 110–13; and rebel cities, 32; and sustainability, 142–5; and sustainable development, 3, 65, 95, 96, 97, 98–100, 109; in Toronto, 138–42, 145–8, 148–53; and transdisciplinarity, 196, 212. *See also* borough governance
Local Government Act (BC), 98, 102

Localism Act (2011), 138
looking, practices of, 200–2. *See also* transdisciplinarity
Los Angeles, 146
Luitzenhiser, Aarin, 180
Luka, Nik, 9, 182, 183
Luxemburg, Rosa, 205

Macdonald, Roderick, 198
Madrid, 30, 33
majoritarianism, 127, 128, 132
Malmö (Sweden), 28
Manaugh, Kevin, 7, 209
mapping, municipal, 96, 104, 106, 107, 108, 109, 110, 111
Markham Street (Toronto), 149, 150, 154
Marx, Karl, 71n4
Masdar (UAE), 17
mastery, 172n18. *See also* experts or professionals
Mazzucato, Mariana, 77n7
McCann, Eugene, 78n8
McGowan, Katharine, 163n1
McMahon, Michael, 4
Meek, Jack, 151
Meijer, Albert, 77n7
Melbourne principles, 22–3
MériteOvation, 68, 74n6, 79
Metcalf Foundation, 153
methane gas, 80. *See also* greenhouse gases
methodologies: and Bloor corridor, 138, 154; and charrettes, 174; and transdisciplinarity, 9, 196–202, 205, 206, 208, 212, 213
Mexico, 31
Miller, Iris, 176, 177, 180, 181n22
Ministry of Transport Quebec (MTQ), 50, 51–2, 53, 59
minor theory, 172. *See also* charrettes, transdisciplinary

Mirvish, Ed, 148
Mirvish redevelopment, 138, 144, 148–53, 154, 156, 157
Mirvish Village BIA, 154
Mirvish Village Task Group, 150–1, 152
Mobilisation Turcot, 51–2. *See also* Turcot Interchange
Mobjörk, Malin, 178
modernization, 17, 23, 27, 31, *36*
Moneris Solutions Corporation, 154
Monforte, Tanya, 9
Montpetit, G., 80
Montreal: and charrettes, 167, 182; and climate action, 194; fossil fuel consultation in, 8, 118, 119, 122, 128–33; green alleys in, 47, 55–8, 59, 84–5; map of, *49*; and participatory governance, 164n3; and Sentier Petit Bourgogne, 53–4; as smart city, 24, 28; sustainable development in, 4, 7, 8, 47, 58–60, 120; and Turcot Interchange, 47–53, 59; as urban commons, 31
Montréal at the Crossroads (Gauthier), 50–1
Montreal Metro, 47–8
Montreal Urban Ecology Centre (MUEC), 182, 183
Moore, Susan, 78n8, 87
Morçöl, Göktug, 140
Morozov, Evgeny, 14, 16, 19, 20, 31, 33, 34
Moses, Robert, 44
Mountain-to-River Project, 47, 59
muddling through, 179, 183
municipalism, 29n2, 30–1, 33
municipalités régionales de comté (MRCs), 82
municipalities, definition, 68–70. *See also* borough governance; innovation; local governance

National Charrette Institute, 168. *See also* charrettes, transdisciplinary
Natural Areas Advisory Committee, 106
neighbourhood associations: and BIAs, 146–7, 148; and Bloor corridor, 154–5, 156; definitions, 145–6; and local governance, 140–1, 142, 147–8; and Mirvish Village, 148–9, 150–3, 157. *See also* stakeholder groups
neighbourhood governance, 138–40. *See also* borough governance
neoliberalism, 17, 18–21, *22*, 26, 28–9, 30, *36*
New Public Management, 67, 77, 86
New Urbanism, 168
New York, 146, 153
Nickles, Thomas, 199
Nishnaabeg, 198–9n2
non-market innovation, 71–4, 75–6, 78, 85. *See also* innovation
non-verbal communication, 180
Norton, Richard K., 112
Novakovic, Stefan, 152

Oddie, Richard, 4
Offe, Claus, 23
Office de consultation publique de Montréal (OCPM), 129, 130, 131, 132–3, 164n3
Okanagan Valley, 97
O'Neill, Bruce, 44
Ontario, 98, 164. *See also* Toronto
"Open Smart City Guide 1.0," 23
Operation Tournesol, 55. *See also* green alleys
Organisation for Economic Cooperation and Development (OECD), 75
originality, 75–6. *See also* innovation
Origins of Totalitarianism, The (Arendt), 201, 203

Orme, Carolee, 152
Osterwalder, Alex, 176, 177, 180
Ostrom, Elinor, 121, 125
Oxford English Dictionary (OED), 168, 169

Palmerston Area Residents Association, 152, 154
Parc des Jazzmen, 54
parking spaces, 8, 55, 58, 100, 154, 155. *See also* vehicles
Park/Pins Overpass, 48
parks, urban: and commons, urban, 121; and innovation, 81; in Montreal, 55; and property rights, 100, 107, 109, 113; and stakeholder groups, 138; and sustainable development, 95; in Toronto, 150, 152, 157
participation, meanings of, 206–9. *See also* collaboration and collaborative governance; stakeholder groups
participation ladder, 20
paved areas, 55, 57. *See also* roads
pedestrian spaces: and Bloor corridor, 154; and green alleys, 55–8; in Montreal context, 47–52, 59–60; and Sentier Petit Bourgogne, 53–4; and sustainable development, 7, 8; and Turcot Interchange, 51, 52–3; urban history of, 44–6
Peñalver, Eduardo, 141
Perng, Sung-Yueh, 19
petitions, 46, 52–3, 56–7, 59, 128–9
Picasso, Pablo, 200
Picketty, Thomas, 213n4
Pigneur, Yves, 176, 177, 180
Place au Soleil, 55. *See also* green alleys
Places to Grow, 98
planning: and charrettes, 9, 140, 163, 166, 167, 169, 171, 176, 180, 182, 183; and green alleys, 56; and municipalities, 68, 78n8, 82, 87; and participatory approaches, 207; and pedestrian spaces, 47, 51; and property rights, 95, 102, 106; and smart cities, 17; and stakeholder groups, 138, 144, 145, 147, 148, 149–51, 152, 157; and sustainability, 142–3; and sustainable development, 99, 101, 125; and transdisciplinarity, 202, 207–8; and transportation, 46, 49–50, 54, 58
Plante, Valérie, 28, 47, 53
Plateau (Montreal borough), 56, 57, 58, 84
platform cooperativism, 32. *See also* rebel cities
policy mobility, 78n8. *See also* innovation; law and policy
political or institutional support: and BIAs, 141, 145; and EDPAs, 96, 102; and green alleys, 54, 55, 56, 57, 58; and neighbourhood associations, 146; and Turcot Interchange, 52, 53, 58
populism, 178
Portney, Kent, 139
poststructuralism, 172n18
power relations: and charrettes, 169, 172, 175–6, 180, 183; and city, definition of, 118n1; and consultation processes, 127, 128, 132–3; and infrastructure, 44; and innovation, 66, 82; and local governance, 53, 139–40; and municipalities, 65, 68–9; and participatory approaches, 206–9; and pedestrian spaces, 46, 47; and rebel cities, 33, *34*, 202; and smart cities, 14–15, 17, 25, 26–7, 28–9, 31, 35; and stakeholder groups,

137, 138, 139–40, 141, 142, 144, 148. *See also* empowerment
praxis, 167, 171
Prince George (BC), 100
principled moral compromise, 174–5. *See also* charrettes, transdisciplinary
private property. *See* property rights
private sector, 24, 28, 67, 76, 86, 180
privatization, 14, 19–20, 109. *See also* property rights
problem-solving: and charrettes, 165, 172, 174, 178; and collaborative governance, 24; and entrepreneurs, 74; and hackathons, 19; and innovation, 72, 86; and smart cities, *36*; and transdisciplinarity, 206, 213
production vs. provision, 125
professionals. *See* experts or professionals
project development, 165, 169, 175, 179, 180
Project Montreal, 51, 53, 57
Promethean response, 17. *See also* smart cities
property rights: in Canada, 97–101; and citizen participation, 109–12; and commons, urban, 121; and EDPAs, 102–4, 105–9; and sustainable development, 8, 95–7, 99–100, 112–13
prototypes and prototyping, 19, 70–1, 72, 73, 75, 175, 176
provision vs. production, 125
Public Consultation Office of Montreal (OCPM), 129, 130
public health, 45, 54, 58, 198
publicity, 79, 127n10, 128, 181
public sector, 67, 72, 76–7, 86, 98–9, 165, 180, 183
purity, 166n8, 206

Quartier 21, 54
Quebec (QC): and climate action, 194; municipal innovation in, 7, 68, 79–84, 87; municipalities in, 65; and sustainable development, 4; and transportation planning, 49–50; and Turcot Interchange, 48–9; and urban commons, 31; zoning in, 128. *See also* Montreal
Québec City, 32
Quebec Ministry of Transport (MTQ), 50, 51–2, 53, 59

Ramadier, Thierry, 172
Rapoport, Amos, 172
rationality, 18–19, 178
readers, 79, 195, 201, 210, 213
rebel cities, 14, 29–35, *36*, 202. *See also* smart cities
Rebel Cities (Harvey), 59
rebound effect, 27. *See also* consumption and consumerism
reciprocity, 120n2
recycling, 80–1. *See also* innovation
"Reduce Dependence on Fossil Fuels in Montreal," 128–30
Red Vienna, 32
Reese, Laura A., 143–4, 156
Reflections on Little Rock (Arendt), 200
reflexivity: and charrettes, 177, 181, 183; and sustainable development, 4; and transdisciplinarity, 9, 198, 201–2, 208, 210, 214
Regional Environmental Council of Montreal (CRE-Mtl), 52–3, 59, 60
regulation: and collaborative governance, 125–6, 127, 128, 131, 132; and commons, urban, 120, 121–2, 130–1; and environmental challenges, 3; and innovation, 66, 73, 76, 80, 81, 82–3, 84, 85, 86, 118–19; and neoliberalism, 18;

regulation (*continued*)
  and pedestrian spaces, 45, 56; and property rights, 8, 96–7, 100–4, 106, 107, 109, 110, 111–12, 113; and stakeholder groups, 141; and sustainable development, 123–4
representation: and charrettes, 169, 173; and collaborative governance, 23, 126, 127; and consultation processes, 106, 128; and innovation, 76; and local governance, 53, 139, 141; and rebel cities, 30, 31; and smart cities, 25, 28; and stakeholder groups, 137–8, 142, 143, 145–7, 150–1, 155–6; and transportation planning, 46
Reseau Express Montreal (REM), 47
resident associations. *See* neighbourhood associations
resilience, 13, 27, 167, 171, 174, 177, 179
resistance: and BIAs, 142; and capitalism, 87; and citizen participation, 99, 125; and innovation, 67; and property rights, 95, 112; and transdisciplinarity, 198, 200–1, 202, 204, 205, 212; and Turcot Interchange, 52, 53. *See also* social movements
resources: and borough governance, 56, 57; and charrettes, 176, 180; and collaboration, 166, 173; and commons, urban, 8, 121–2, 124, 125; and consultation processes, 119–20, 128, 130–1, 132; and innovation, 66, 70, 71–4, 76, 78, 79, 83, 84, 85; and local governance, 139, 140–1, 144, 146; and mapping, 104; and Nishnaabeg ethics, 198–9n2; and smart cities,

13, 16, 27, 28, *36*; and sustainable development, 111, 143, 194
restoration, 22, 96, 97, 102–3, 108, 109, 111, 112
Rieti, John, 18n1
right to the city, 34, 46. *See also* inclusion and inclusivity
Rio Earth Summit (1992), 21
riparian areas, 96, 101, 102–3, 104, 113
Riparian Areas Protection Act, 103
Riparian Areas Regulation (RAR), 104
Rittel, Horst W. J., 163n1
roads, 44, 45–6, 50, 79, 97, 130. *See also* Turcot Interchange; vehicles
Robinson, Pamela, 4
Rogers, Dennis, 44, 54
Roggema, Rob, 170, 176–7
Rojava revolution, 30. *See also* rebel cities
Roseland, Mark, 97
Rosemont-La-Petite-Patrie (Montreal borough), 84–5

Saanich, District of, 95–6, 104–9, 110–12, 113
Saanich Citizens for a Responsible EDPA, 105–6, 107
Sabel, Charles, 119
Sacerdote, Bruce, 212
safe spaces, 177, 179, 183. *See also* charrettes, transdisciplinary
Saint-Hyacinthe (QC), 80
Sallaberry (QC), 81–2
sanctuary cities, 32. *See also* rebel cities
Sassen, Saskia, 3, 199–200, 201, 203, 205, 208–9, 211
Sauter, Molly, 18
Schleicher, David, 196, 212, 213n4
Schneider, Nathan, 32
Scholz, Trebor, 32
Schumpeter, Joseph, 70, 71n5

Schumpeterian innovation, 68, 70–2, 73
Seaton Village Residents' Association, 149
Sensitive Ecosystems Atlas, 104
Sentier Petit Bourgogne, 47, *49*, 53–4, 59
Sharing Cities Seoul Initiative, 32
Shearmur, Richard, 7, 74n6, 77n7, 203
Shotwell, Alexis, 166n8
Sidewalk Labs, 18. *See also* Toronto
Simon Fraser University, 99
simple language, 211. *See also* transdisciplinarity
Simpson, Leanne, 198–9n2
situatedness, 176, 178, 179, 180, 181, 209
Slack, Enid, 97–8
Sleiman, Mark, 142
smart cities: citizen-centred rebel model, 29–35; and citizen participation, *22, 26, 34*; collaborative sustainable model, 21–9; definitions and context, 13–16; and innovation, 77; models of, *36*; and sustainable development, 6–7; techno-centric neoliberal, 17–21
smart growth, 15, 97–8, 99, 169n12. *See also* growth management
snow removal, 83
social capital, 29, 88, 119
social liberalism, 25, 28–9, *36*
social monitoring, 25, 33
social movements: and charrettes, 167, 182; and climate change, 194–5; in Montreal, 50–3, 54, 56–7, 128–9; and municipal innovations, 87; and property rights, 95–6; and rebel cities, 29–32, 33; and sustainable development, 4, 5, 6–7; in Toronto, 139, 144, 155–6; and

transdisciplinarity, 200, 209, 211. *See also* resistance
Soja, Edward, 46–7
solidarity, 32, 52, 141
Songdo International Business District (South Korea), 17–18
sovereignty, technological, 31–2, 34, *36*. *See also* smart cities
Spadina Expressway, 137, 139
Spain, 30–1
specialists. *See* experts or professionals
sprawl, urban, 97–8, 123, 124
stakeholder groups: and Bloor corridor, 153–7; and charrettes, 164, 166, 168, 171, 173–7, 178, 181, 182–3; and democratic processes, 8, 127, 128; meanings of, 195; and Mirvish Village, 148–53, 156; and smart cities, 24, 25; and sustainability, 142–5; and Toronto local governance, 8–9, 137–42, 146–7; and transdisciplinarity, 207, 208–9. *See also* borough governance; collaboration and collaborative governance; consultations
Ste-Catherine (Montreal), 47, 59–60
St-Lawrence Seaway, 48
subjectivity, 18–19
sustainability: in Canada, 194; and charrettes, 169, 174, 177, 178, 182; and city governance, 3; and collective action, 166n8; definitions, 21, 22, 78, 122–3, 142–3, 195; and local governance, 139, 142–5; and Mirvish Village, 148, 153; and municipal innovation, 65, 68, 71, 78–85, 87–8; and rebel cities, 32–3, 34–5; and smart cities,

sustainability (*continued*)
13, 14–15, 15–16, 21–6, 26–9, *36*;
and social liberalism, 28–9; and
stakeholder groups, 137, 138, 157;
and transdisciplinarity, 200, 204–5,
209, 212; and treaty relationships,
198–9n2
sustainable development, definitions,
3–4, 122–3, 143
sustainable urban development:
approaches and literature, 3, 4, 5,
6; and charrettes, 9, 163–83; and
consultation processes, 8, 118–34;
and coordination, challenges of,
5–6; definitions, 122–4; and law,
discipline of, 4–5; and municipal
innovation, 7, 65–88; and pedestrian
space, 7, 44–60; and property rights,
8, 95–113; and smart cities, 6–7,
13–36; and stakeholder groups, 8–9,
137–57; and transdisciplinarity, 9,
183–214
Swann, Peter, 78n9, 85
Syrian Kurdistan, 30

tar sands, 85. *See also* innovation
taxes, 70, 83, 100, 110, 148
techno-centrism, 17–21, *22*, 26, 27, *36*
Temenos, Cristina, 78n8
Termeer, Catrien J. A. M., 163n1
Terrebonne (QC), 81–2
Thaens, Marcel, 77n7
Third Way, 28–9
Thompson, Dennis, 120n2, 127n10
Thompson, Ken, 139
Thompson, Mary A., 176
Thrift, Nigel, 13
Tiebout, Charles, 69–70, 125
tokenism, 25, *26*, 46, 169n13, 208
Toker, Umut, 173
Toronto: and Bloor corridor, 153–7;
and Mirvish Village, 148–53; and
Sidewalk Labs, 18; stakeholder
groups in, 140, 145–8; sustainable
development in, 8–9, 137–8, 143
Toronto Centre for Active
Transportation, 153
Toronto Dollar, 32
totalitarianism, 201–2, 203
tradition, 66. *See also* innovation
Trans-Canada Highway, 48. *See also*
roads
transdisciplinarity (TD): and
accessible language, 210–12;
and expertise, 202–4; five
characteristics of, 200; importance
of, 212–14; and inclusivity, 178n21,
206–9, 213; and innovation,
198–200; and knowledge, 204–6;
and looking, practices of, 200–2;
methodological approach of, 9,
194–8, 199. *See also* charrettes,
transdisciplinary
TransformTO, 153
Transport Actif Parc-Extension, 59
transportation: and Bloor corridor, 153,
154; and collective action problems,
8; and infrastructural violence, 44–5,
46; in Montreal, 7, 47–52, 53–4, 58,
59, 131, 182; and smart cities, 15,
20, 32; and stakeholder groups, 144;
and sustainable development, 98, 99,
112–13
treaty relationships, 198–9n2
Tremblay, Gerard, 51
Trois-Rivières (QC), 83
trust, 108, 125, 131, 172, 177, 178
Turcot Interchange, 47–53, 59

Uber, 14, 18, 19–20, 32, 76
undisciplined eyes, 200–2. *See also*
transdisciplinarity
Union des municipalités du Québec
(UMQ), 68, 83, 87

United Kingdom, 138
United Nations, 54, 128–9
United States, 31, 32, 45, 99, 100, 101, 168
unity, 172
universities: and charrettes, 181, 183; and collaborative governance, 24, 25; and consultation processes, 131, 153; and knowledge production, 208, 209; and sustainable development, 98, 99; and Turcot Interchange, 51
upland ecosystems, 104
urban commons, 31–2, 118, 119–22, 122n4, 124–6, 130–2, 133–4
urban design, 9, 18, 164n3, 167, 169, 176. *See also* collaboration and collaborative governance; urban commons
urban development: environmental impact of, 123; and neoliberalism, 19–20; and pedestrian spaces, 45; and rebel cities, 31, 33; and smart cities, 13, 14–15, 17, 18, 21, 23, 27, 28–9, 35; and smart growth, 98. *See also* sustainability
urban governance. *See* borough governance; collaboration and collaborative governance; local governance
urban studies, 4, 167, 195, 200, 207, 212
Urban Sustainable Directors Network, 65
utilitarianism, 141

value (economic): and innovation, 66, 72, 73, 75, 81–2, 83, 85–6, 203; and property rights, 100–1, 105, 106–7, 108, 109, 110, 111, 112
Vancouver (BC), 97, 98, 99, 112, 143, 149

Van Duzer, C. H., 203
Vasavada, Triparna, 140
vehicles: and Bloor corridor, 154; and collective action, 8; and commons, urban, 130–1; impacts of, 44–5, 123, 124; in Montreal, 48–9, 50, 51, 55, 56, 57, 58; and right to the city, 46; in Saint-Hyacinthe, 80; in Trois-Rivières, 83
Victoria (BC), 99, 100, 112, 124, 125
Ville-Marie Expressway, 54
visual thinking, 180
vocabularies, 174, 195–7, 199, 202, 203, 204, 210, 212–13

Wagenaar, Hendrik, 86
walking: and charrettes, 179; in Montreal, 44–6, 47, 54, 58, 59, 182; in Toronto, 143, 153
Wang, Xiao Hu, 125
Ward, Kevin, 151
Warren, Robert, 125
Washington, 142
waste and waste management, 15, 17, 32, 68, 76, 77, 78, 80–1
water and water management: in BC, 99, 113; and cities, 3; and EDPAs, 97, 102, 103, 111, 112; and innovation, 68, 76, 81–3, 86; in Montreal, 55; and sustainable development, 123; and technological sovereignty, 32; in Toronto, 143
Wates, Nick, 173
Webber, Melvin M., 163n1
Weinstock, Daniel, 174–5
Westbank (developer), 148, 151, 152
Westley, Frances, 163n1
wicked problems, 163, 164, 178, 182
Wiesman, Urs, 208
Wilcox, David, 35
Wilkinson Road (Saanich), 104

Wills, Jane, 79, 87
Wolf, James, 142
Wood, Matt, 86
Woodward's building, 149
workshops, 9, 167, 169, 172, 173, 176, 182, 213. *See also* charrettes, transdisciplinary
World Commission on Environment and Development (WCED), 3, 21
World Expo 1967, 47
Wright, Erik Olin, 139

Yamaska municipal region, 82
Young, Oran, 87

zoning: in BC, 97, 99, 124; and charrettes, 140; and collaborative governance, 127–8, 132; and DPAs, 96, 102, 103–4, 111; and Mirvish Village, 149, 151, 152; and property rights, 100–1; and sustainable development, 143; and urban sprawl, 123; and vehicles, 45

www.ingramcontent.com/pod-product-compliance
Lightning Source LLC
Chambersburg PA
CBHW020251030426
42336CB00010B/711